A Participatory Approach to
MODERN GEOMETRY

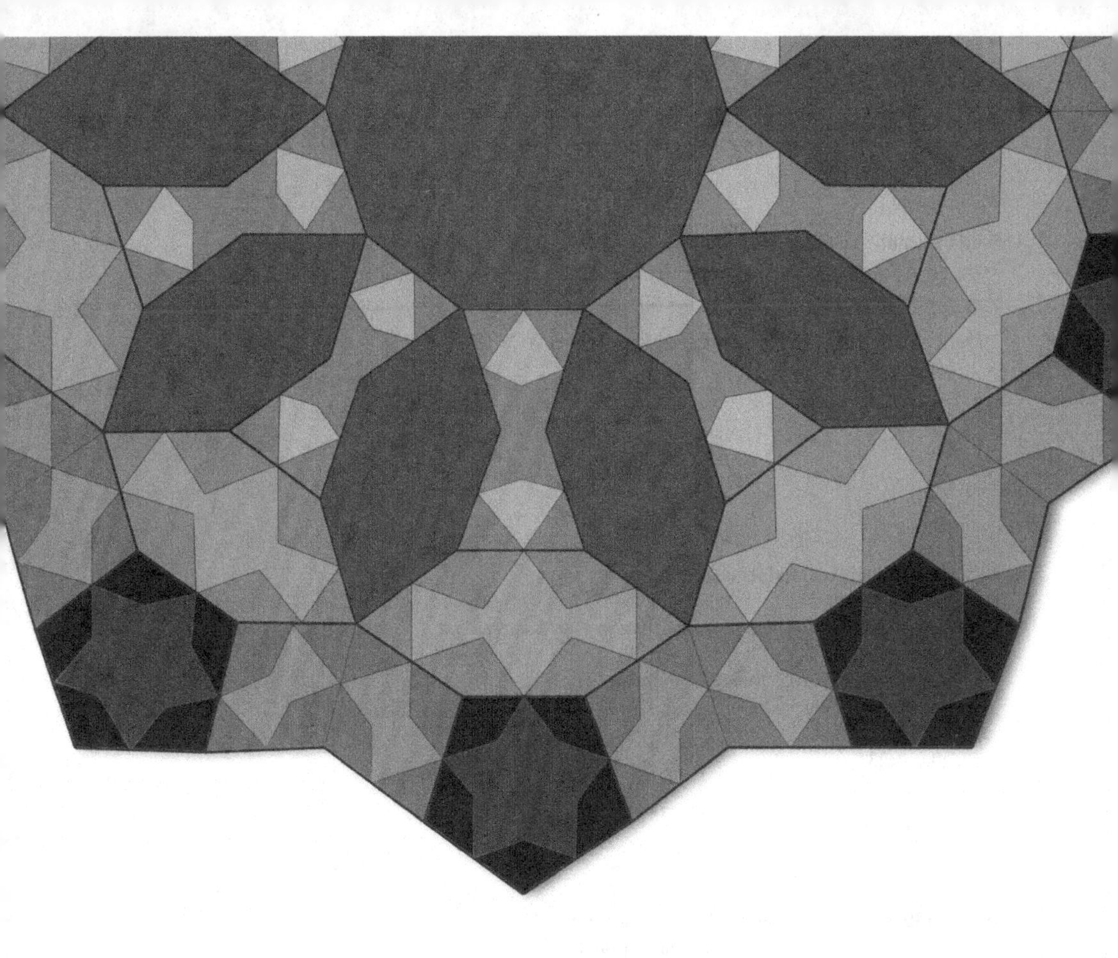

A Participatory Approach to MODERN GEOMETRY

Jay Kappraff

New Jersey Institute of Technology, USA

NEW JERSEY • LONDON • SINGAPORE • BEIJING • SHANGHAI • HONG KONG • TAIPEI • CHENNAI

Published by

World Scientific Publishing Co. Pte. Ltd.
5 Toh Tuck Link, Singapore 596224
USA office: 27 Warren Street, Suite 401-402, Hackensack, NJ 07601
UK office: 57 Shelton Street, Covent Garden, London WC2H 9HE

Library of Congress Cataloging-in-Publication Data
Kappraff, Jay, author.
A participatory approach to modern geometry / Jay Kappraff.
pages cm
ISBN 978-9814556705 (hardcover : alk. paper)
1. Geometry. I. Title.
QA445.K3574 2014
516'.04--dc23

2013035770

British Library Cataloguing-in-Publication Data
A catalogue record for this book is available from the British Library.

Typeset by Stallion Press
Email: enquiries@stallionpress.com

Printed in Singapore

CONTENTS

PREFACE

The study of geometry goes back to the beginning of civilization. Perhaps the first recorded book of geometry was the Sulba Sutra written around 600 B.C. It contained the famous Pythagorean Theorem. It is said that at the entry to Plato's Academy were the words, "Let all who wish to study geometry enter here." The Greek Quadrivium stated that a student should keep his eye on unity and study: Geometry, Astronomy, Music and Number. Although there is strong evidence that Egyptians and Babylonians were doing geometry long before the Greeks, the Greek mathematicians, and in particular Euclid, were responsible for placing this subject on a logical foundation which became the gold standard for validating mathematical and scientific truths.

In this book, we will pay homage to the great traditions of geometry by introducing the art of proof. However, we will not dwell on the proofs. Instead, we will focus on how geometry can be applied in meaningful ways and how it gives you a power to move towards interesting discoveries of modern mathematics.

Traditionally, geometry was carried out using crude instruments such as a compass and straightedge. In fact there is evidence that these tools were used to construct the great architectural structures of antiquity and study the movements of planets and stars. We will follow this path. However, with the advent of the computer, geometrical constructions can now be implemented with greater speed and precision. So after first encountering the subject with compass and straightedge, I invite you to use drawing programs such as Corel Draw, Adobe Illustrator, GeoGebra or Photoshop to carry out the constructions.

In line with the title of the book, *A Participatory Approach to Modern Geometry*, the book emphasizes hands-on experience. Throughout the book you will find invitations to participate in the narrative. So you should read the book with the instruments of geometry close at hand. Several of the topics suggest designs that you can carry out pertaining to geometry. You can do this by hand or using computer-drawing programs. The book has twenty-four chapters and each chapter has both explanatory material, problems, and constructions to challenge you.

This course will focus primarily on Euclidean geometry based on Euclid's thirteen books. However, in order to better appreciate this subject, an introduction to projective geometry will be provided in Appendix A. However, we will not present the theory of this subject, but rather a sequence of thirteen projective geometry constructions with no explanation. The subject of projective geometry is familiar to students of architecture and design through its application to perspective. It makes the notion of infinity tangible since the horizon in a scene is rendered by a horizontal line on the canvas. Also, while in Euclidean geometry two lines intersect at a point except when the lines are parallel, in projective geometry every pair of lines intersect; when the lines are parallel they intersect at infinity, i.e., a point on the horizon.

In general, mathematics is based on the idea of transformation. An object is transformed by some rule and the question is asked: What remains unchanged? The transformations pertaining to Euclidean geometry are rigid body motions in which length, area and angle and, more generally, the distance between any pair of points remain unchanged. This leads to the idea that two objects are equivalent (in Euclidean geometry, congruent) if they can be moved and matched point for point with each other or their mirror images. In the context of projective geometry, two objects are equivalent if one can be matched point for point with the other by projection from a point or a series of points. What remains constant under these circumstances is much more subtle. Angle, length and area change but one rather abstract quantity called cross-ratio remains unchanged. This will be described in Construction 1 of Appendix A. In fact, it can be shown that under the transformations of projective geometry, all triangles are equivalent as are all conic sections: circles, ellipses, parabolas and hyperbolas. Projective transformations also shows that space is not free-wheeling, but geometry imposes certain constraints on space.

In Chaps. 1–15, we study the standard elements of geometry such as the Pythagorean Theorem, angle, perpendicular and parallel lines, area, congruence, similarity, compass and straightedge constructions and the theorems that validate these constructions. After introducing the construction of triangles using compass, straightedge, and protractor, given partial information about the triangles, we show how trigonometry can determine sides and angles of a triangle without the need to construct. In other words, if you can construct a triangle from partial information with compass, straightedge and protractor, then you can also solve that triangle by trigonometry. Chapter 16 is devoted to a study of logarithmic spirals, showing their origin to be a simple right triangle. Spirals will lead to a discussion of an ancient technique referred to by Jay Hambridge as dynamic symmetry. This will lead to a discussion of the golden and silver means and their application to design in Chap. 17. Beginning with Chap. 18, we place the study of plane Euclidean geometry squarely in the study of transformational geometry and isometries: translations, rotations, reflections and glide reflections. We introduce the notion of an isometry and apply it to kaleidoscopes, an introduction to group theory and its application to kaleidoscope and frieze symmetry. Chapter 23 is devoted to fractals while Chap. 24 employs matrices to carry out rotations, reflections, translations and projective transformations.

The book has a number of novel topics not found in other courses of geometry. Chapter 1 introduces the study of geometry by construction of grids formed by the intersection of circles in a manner that goes back to the origins of geometry in ancient civilizations. The Pythagorean Theorem is presented through a set of imaginative proofs involving the student in hands-on experiences. Chapter 3 applies the theory of lines to scan-converting a line in terms of pixels. Chapter 6 presents an extensive introduction to trigonometry. Chapter 9 applies perpendicular lines to a study of Voronoi domains. Chapter 10 presents an introduction to the Poincare plane as a means of studying non-Euclidean geometry. Chapter 11 applies parallel lines to the bracing of a framework. Chapters 14 and 15 develop the concept of area through the use of geoboards and vectors. Transformational geometry leads to experiments with kaleidoscopes in Chap. 20, an introduction to group theory in Chap. 21, and a study of fractals in Chap. 23 where the students engage in creating a giant fractal wallhanging.

Although proof is not emphasized in the course, geometry is developed in a systematic and rigorous manner. Many proofs are given, and students are asked to provide others. Before engaging in a proof, the book often reveals first the strategy behind the proof before showing its details.

A word about the level of this course. The book was written to be understood at the freshman level by students of architecture and design although it is suitable for students in the higher levels. This book was written because most books of geometry on the college level are written for upper class mathematics majors. However, the constructive approach and the novel material in the book should make it of interest to math majors and engineering and science students who wish to improve their visualization skills. I have had the experience of two upper level mathematics students taking the course. They claimed to derive great benefits from this approach.

Finally, the notation of geometry can be very confusing with different symbols for line segments and angles as distinguished from the measure of these line segments and angles. Also lines can be directed or not. Sometimes we mean our notation to pertain to an entire line or a line segment or a half line or ray. In this book, we have chosen not to be rigorous about the notation and leave it to the context as to what we mean in any instance. Also, geometry is heavily laden with terminology. As an aid to the student, we provide a glossary of geometry terms found throughout the book in Appendix B.

There are many routes through this book. The entire book can be comfortably taught in a one-semester four-credit course. I generally introduce one projective construction each week and thus cover all thirteen constructions over the course of the semester. The constructions are self-explanatory, so a five-minute introduction is usually sufficient. If you only have a three-credit course like I do, you can omit Chap. 24 and abbreviate other chapters. There is also the option not to do the projective constructions from Appendix A or assign them for extra credit. Depending on the interests and level of the students, you can choose to eliminate Chap. 6 on trigonometry. It has been included because I have found students lacking in trigonometry skills. This topic is particularly important for students of architecture. I generally give a midterm and a final exam. However, I have found that some students who have an excellent visual understanding of the material have poor computational skills. For such students, they are able to compensate for poor exam grades with well-executed constructions and auxiliary designs suggested in several of the chapters. I also assign essays requiring the students to reflect on the nature of geometry. Several essay suggestions are given in Appendix C.

In keeping with the title of the book, *A Participatory Approach to Modern Geometry*, the student is asked to engage in constructive activities throughout the book. The student should bring to class on a daily basis a toolkit with the following items: compass, ruler (straightedge), a 30, 60, 90-triangle, protractor, some markers, graph paper, glue stick and scissors. I have found circle-master compasses to be of the best quality and generally supply these to the students. The teacher should also supply the students with rectangular mirrors for Chaps. 20 and 21, soap erasers and ink pads for Chap. 22, and construction paper for Chap. 12. The instructor will also need, for demonstration purposes, chalkboard instruments: compass, straightedge, protractor and 30, 60, 90-triangle. The above items are available on the web from the following suppliers.

Circle-master compasses — Nasco

Mirrors — EAI Eduction

Stamp pads — Staples

Soap erasers — Jerry's Art Supplies

Chalkboard instruments — Barclay School Supplies.

You will notice a design which is placed at the beginning of each chapter of this book created by Teja Krasek, a Slovenian computer artist. This design illustrates several of the themes that you will find throughout the book. It exhibits the power of the golden mean as a design tool. The design incorporates pentagons, decagons, golden triangles, and golden diamonds in a fractal like structure in which its elements vanish towards infinity. The design also has a symmetric motif which gives the illusion of rotation about a center. This book will work toward clarifying the notions inherent in this design as the course; progresses.

A teachers' manual is in preparation which will contain helpful hints on how to use this book to teach a course in modern geometry. It will contain solutions to problems and additional problems pertaining to each chapter.

Acknowledgement

I wish to acknowledge the help of Kevin Miranda who created most of the figures that you will find throughout the book and added a measure of order to the process of creating this text. Branko Grunbaum and Denis Blackmore were helpful in reading the book as it was being created and making valuable comments. Any errors that are found in the book are my responsibility. Teja Krasek created the design which introduces each chapter and several other figures found in the book. World Scientific, and in particular my editor, Tan Rok Ting, helped to facilitate my vision of a student-friendly textbook on geometry. Others who contributed to the figures found throughout the book are Mie Shimobayashi, Vania Masilang, and Hsin Ting Hsieh. Thanks also for Cindy Caldas, Joseph Titterton, Elliot Virtgaym, and Mark Bak who contributed designs to the book. A special thanks to Tatiana Garay who created the lovely design for the cover of the book. Most of all I wish to thank the many students who have taken this course in the earlier stages of its development, and for their helpful comments.

Jay Kappraff
September 2014

CHAPTER 1

TRIANGLE AND SQUARE GRIDS

1.1. Introduction

The power and potential that exists by the simple act of drawing a circle or a square is amazing. This was recognized in ancient times where the circle symbolized the celestial sphere and represented the Heavenly domain; the square, with its reference to the four directions of the compass, represented the Earthly domain. Margit Echols [Ech] was a quilter who used geometry as the basis of her wonderful quilt patterns. In this chapter, Echols shows how to create triangle and square grids beginning with a circle. The grids then become scaffolding for the creation of a limitless number of repeating patterns. These patterns are sometimes referred to by sacred geometers as The Flower of Life [Flo], [Kap2].

1.2. Triangle-Circle and Triangle Grids

For this exercise, you may use either a compass or a software program such as Corel Draw, Adobe Illustrator, or Geometer's Sketchpad to create circles. To create a triangle grid, start with a circle. Place the compass point on the circumference and draw another circle with the same radius. The resulting pair of circles is perhaps the most fundamental structure in geometry called the *Vesica Pisces* shown in Fig. 1.1a. This figure had sacred significance in the Christian religion. It is found in many churches where images of Christ were drawn in the central fish-like region as shown in Fig. 1.1b. The Vesica Pisces is created by starting with a line and a single point on the line which is the center of a circle as shown in Fig. 1.2. Where the line intersects the circle, draw another circle of the same radius. The Vesica Pisces is the starting point of a triangle-circle grid. Figure 1.3 shows how to generate such a grid step by step. By Step 9, you have created a triangle grid. Look within the grid. Can you see a grid of hexagons embedded in the grid of equilateral triangles? Islamic artists designed

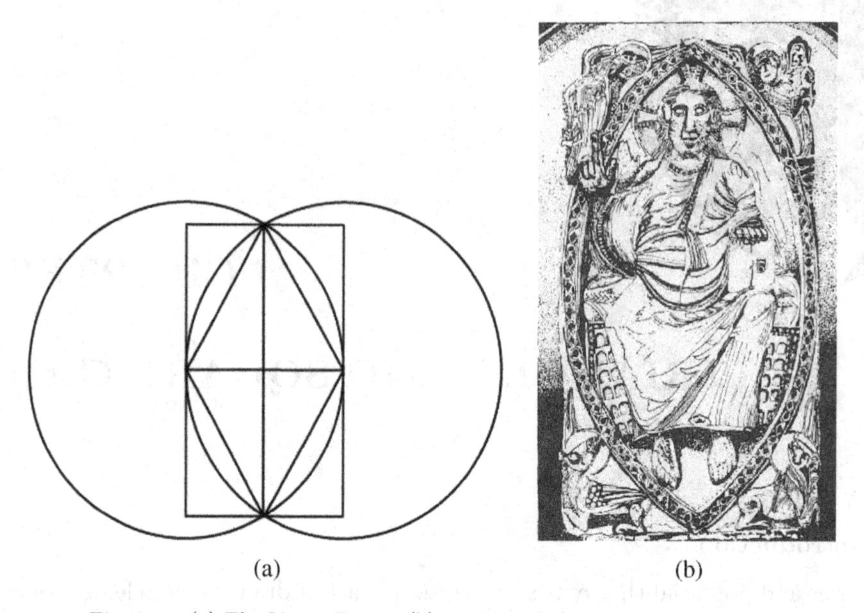

(a) (b)

Fig. 1.1. (a) The Vesica Pisces; (b) marble relief of Christ in a Vesica.

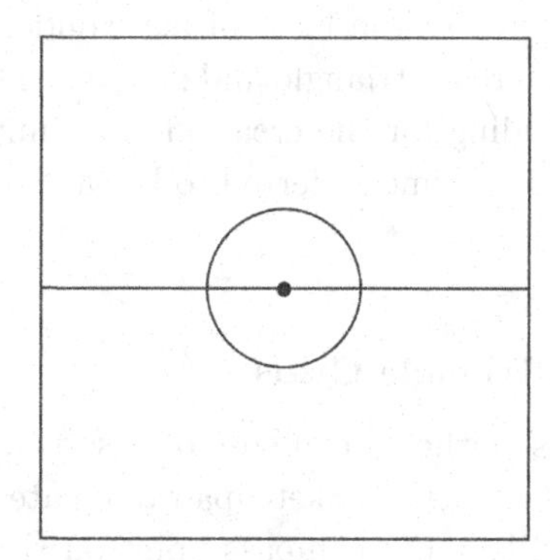

Fig. 1.2. Geometry begins with a point, a line and a circle.

patterns with hexagons and 6-pointed stars on this grid from the center outward with no gaps and no overlaps as seen in the shaded stars of the triangular grid in Fig. 1.4.

Pattern Puzzle A is presented in Fig. 1.5a. These patterns are hidden in the triangle-circle system of Fig. 1.5b. Draw each of these patterns using the overlapping circles in Fig. 1.6b as a guide. You can add color. Note that some patterns can be traced directly from the curved lines, while the triangle grid can be drawn using a straightedge to connect points where the circles intersect. Many other patterns can be

Drawing Circle and Triangle Grids

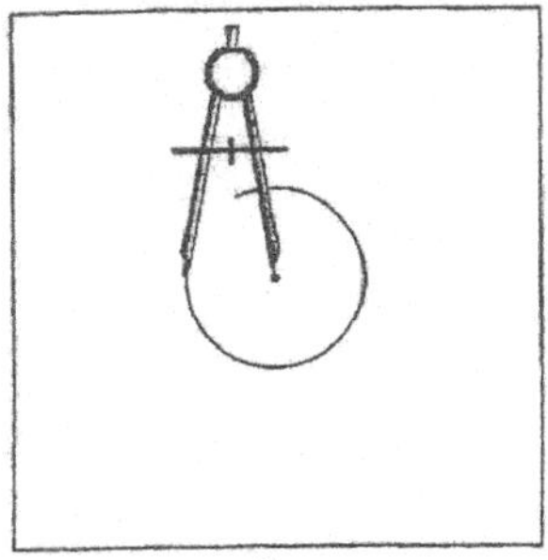

1. Draw a circle.

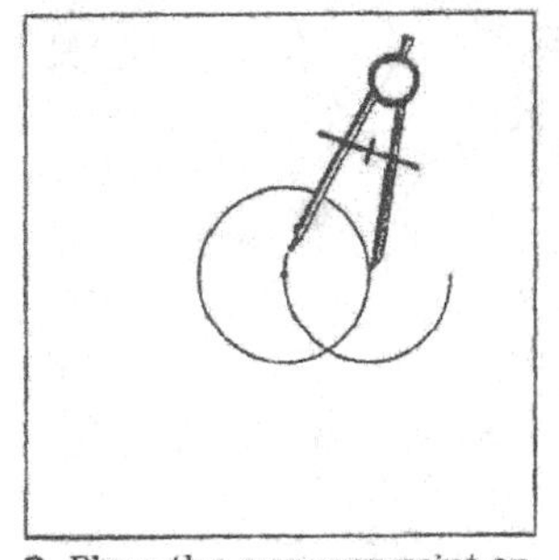

2. Place the compass point on the circumference and draw another circle.

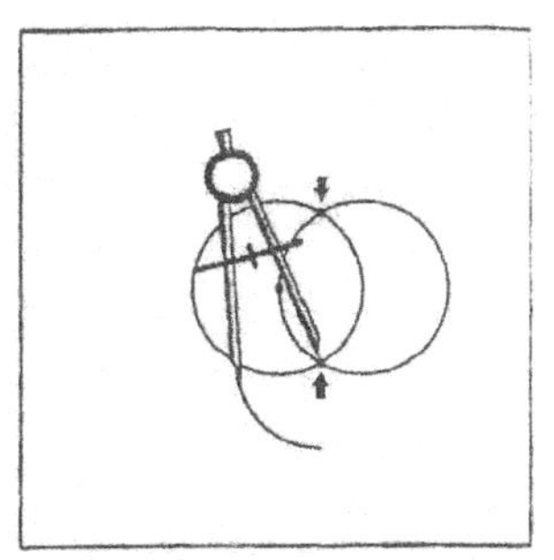

3. Place the compass point at points of intersection and draw 2 more circles.

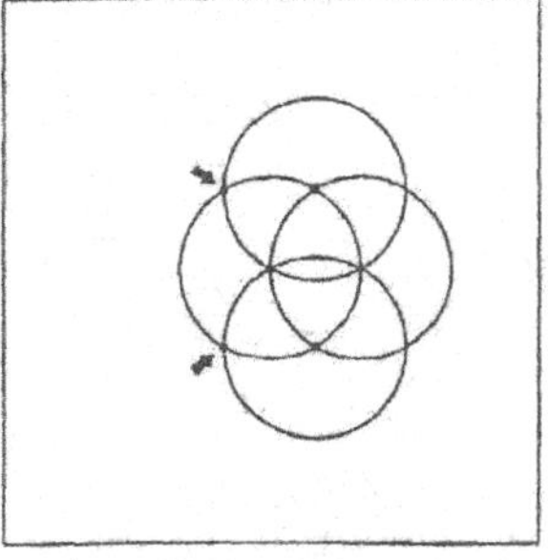

4. Place the compass point at new points of intersection and draw 2 more circles.

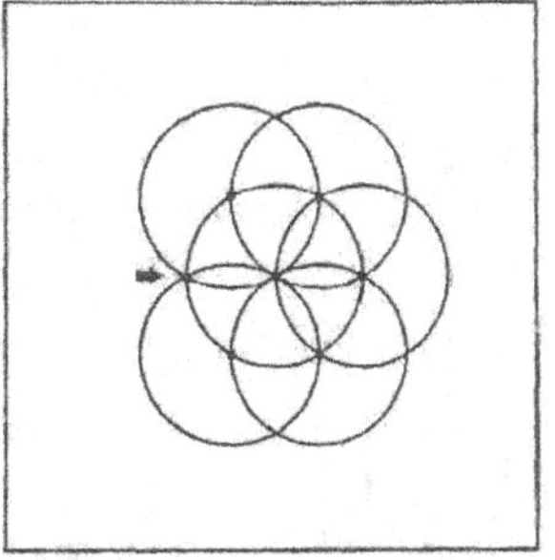

5. Place the compass point at the last intersection on the first circle and draw a circle.

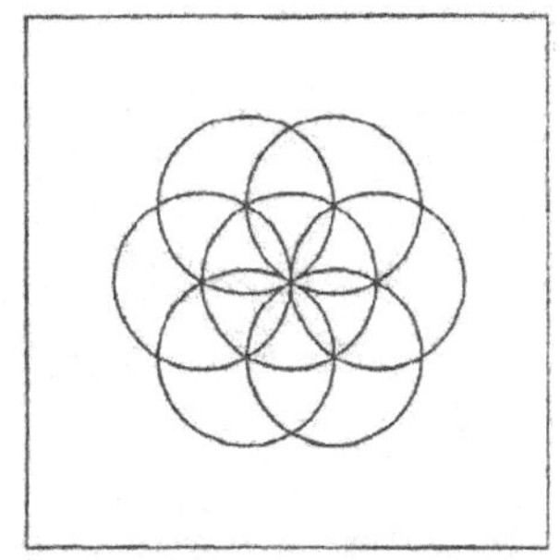

6. The circles divide the first circle into 6 equal parts.

7. To make a circle grid, continue adding circles until the page is filled.

8. Connect the centers of the circles to create a grid of equilateral triangles.

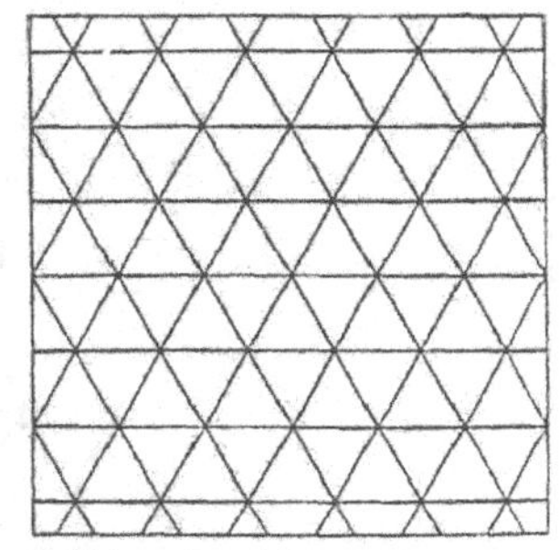

9. Triangle grid.

Fig. 1.3. Nine steps to drawing a triangle-circle grid.

found in this triangle, such as the circle grid. Try to find others, or invent some of your own. The possibilities are endless!

Pattern Puzzle B is presented in Fig. 1.6a. Look for these patterns hidden in the triangle grid in Fig. 1.6b. You can either shade or add color. You can also draw each one on a sheet of tracing paper using the underlying grid as a guide or use your computer

Islamic artists designed patterns
on the grid from center outward with
no gaps and no overlaps.

Fig. 1.4. Islamic artists designed star and hexagonal patterns on a triangular grid.

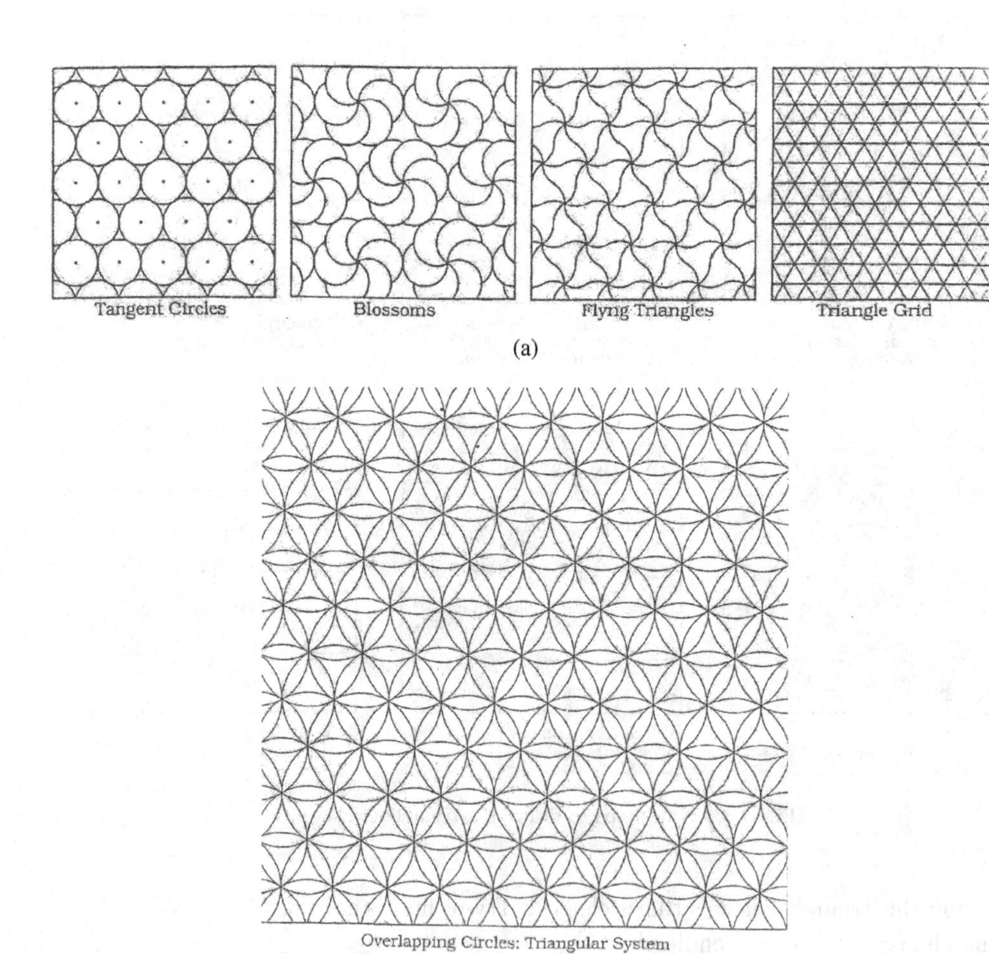

Fig. 1.5. (a) Pattern puzzle A. Can you find these patterns in the triangle-circle grid?; (b) A triangle-circle grid.

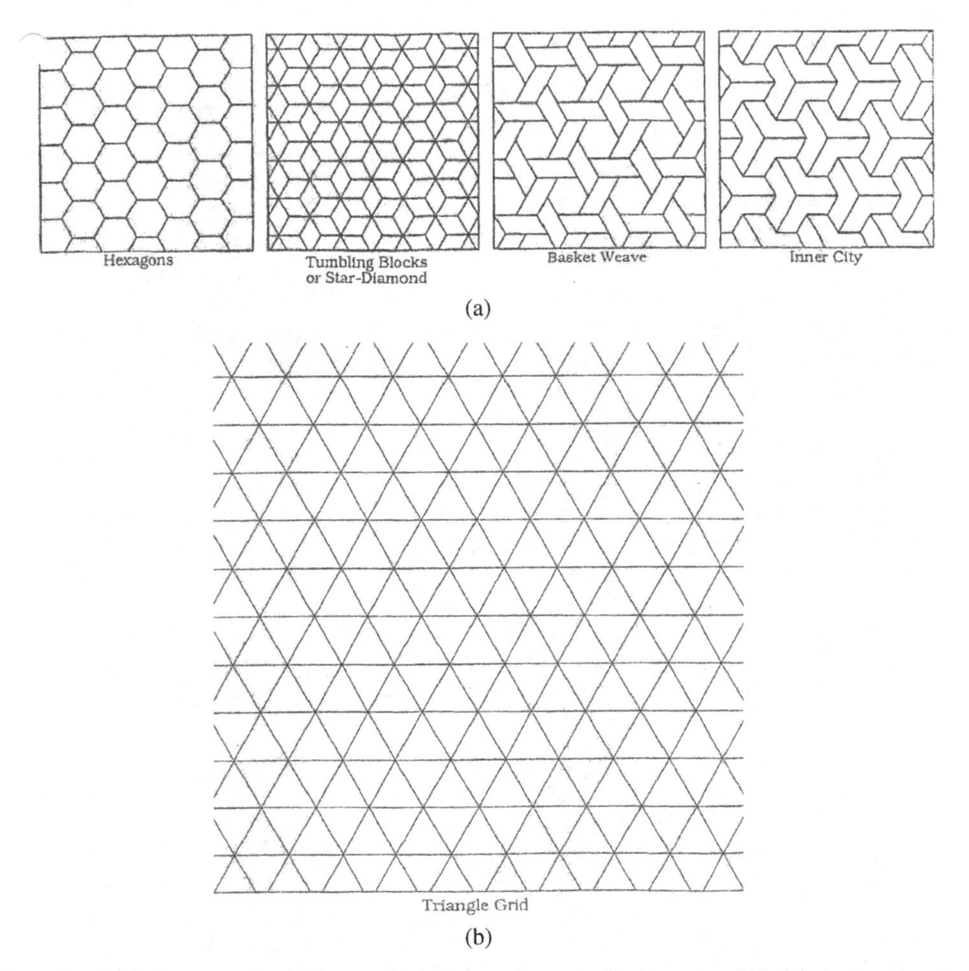

Fig. 1.6. (a) Pattern puzzle B. Can you find these patterns in the triangle grid?; (b) A triangle grid.

drawing programs. You may try creating patterns of your own. Again, there are an infinite number of possibilities.

1.3. Square-Circle and Square Grids

Whereas the triangle-circle grids were created by beginning with a line and a single point on it, square-circle patterns begin with a pair of lines intersecting at a right angle. The point of intersection of these lines defines the center of a circle. A circle is drawn about this center in order to define the four points where the pair of perpendicular lines intersect the circle. The initial circle is then removed. The four new points are centers of new circles with the same radius. The points where two circles intersect define the centers of new circles of the square-circle grid. Construct a square-circle grid by following the procedure in Fig. 1.7 step by step. This grid again leads to countless designs, ten of which are shown in Pattern Puzzle C in Fig. 1.8.

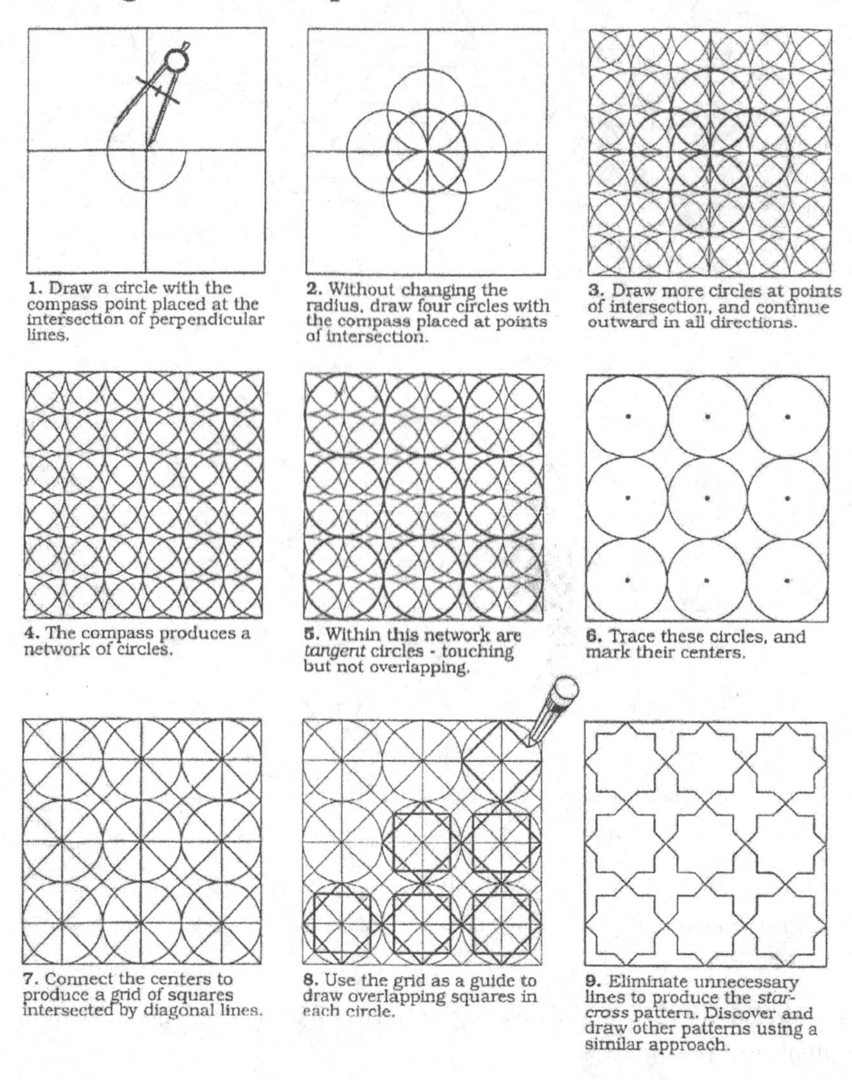

Drawing Circle and Square Grids, and Star-Cross Pattern
1. Draw a circle with the compass point placed at the intersection of perpendicular lines.
2. Without changing the radius, draw four circles with the compass placed at points of intersection.
3. Draw more circles at points of intersection, and continue outward in all directions.
4. The compass produces a network of circles.
5. Within this network are tangent circles - touching but not overlapping.
6. Trace these circles, and mark their centers.
7. Connect the centers to produce a grid of squares intersected by diagonal lines.
8. Use the grid as a guide to draw overlapping squares in each circle.
9. Eliminate unnecessary lines to produce the star-cross pattern. Discover and draw other patterns using a similar approach.

Fig. 1.7. Nine steps to drawing a circle and square grid and the star-cross pattern.

1.4. Star Designs Based on the Triangle Grid

Beginning with the triangle-circle grid as shown in Fig. 1.9, six- and twelve-pointed stars can be constructed. For this exercise you can also use a coffee can cover instead of a compass to carry out the construction in which the center of the cover has the usual marking; you simply place the center of the coffee can cover over the point that you would otherwise place your compass point to draw the circle. Beginning with the square-circle grid, eight- and sixteen-pointed stars can be created as shown in Fig. 1.10.

Fig. 1.8. Pattern puzzle C. Can you find these patterns in the square-circle grid?

Drawing Six- and Twelve-Pointed Stars

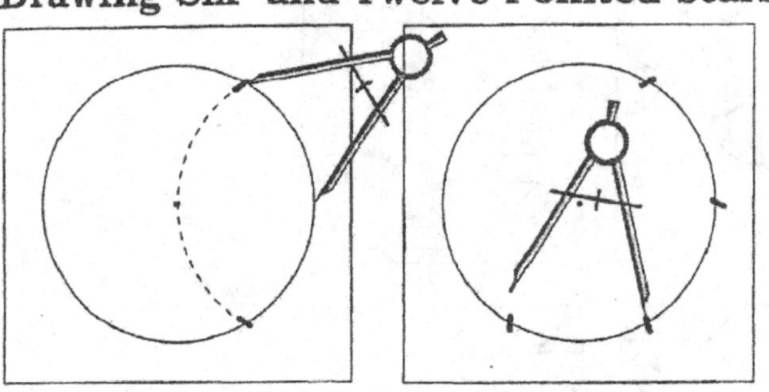

1. Draw a circle. Without changing the radius, place the compass point on the circumference, swing it to each side, and make two marks. These marks *intersect* the circle.

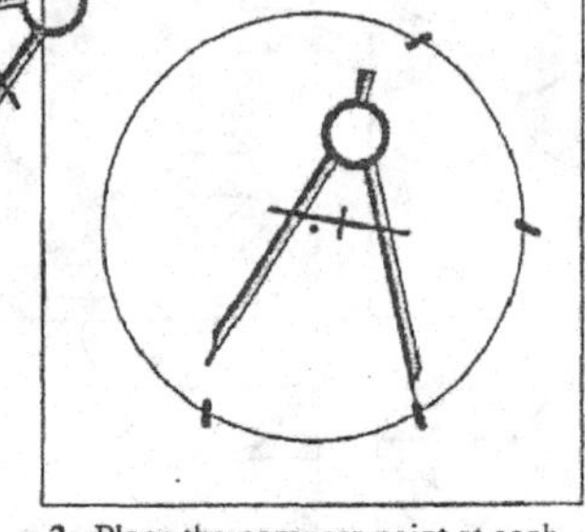

2. Place the compass point at each intersection, swing the compass to each side as before and make two more marks. Repeat until the circle is divided into six equal parts.

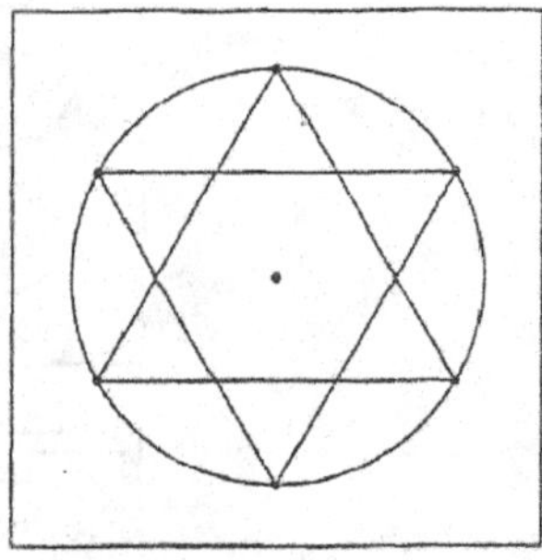

3. Connect every other point to make a six-pointed star. Note the hexagon in the center. How many equilateral triangles does the star contain?

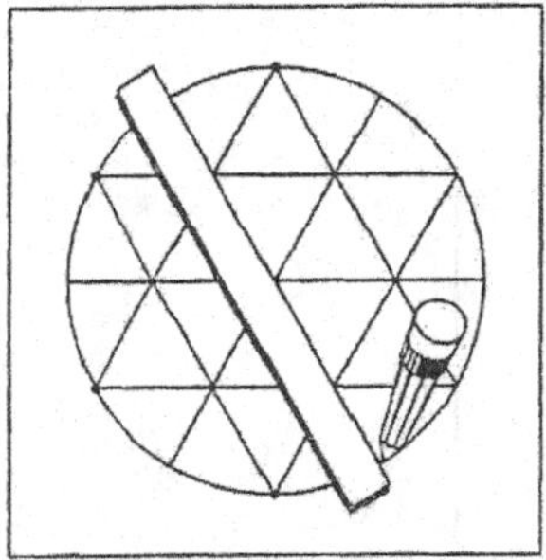

4. Draw lines through the center of the star and opposite vertices of the hexagon.

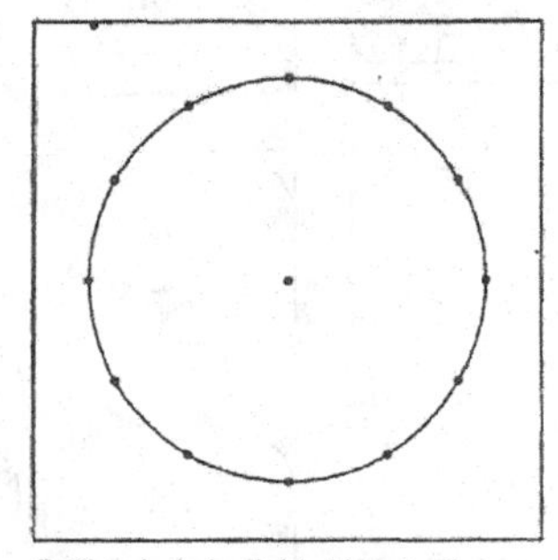

5. The circle is divided into twelve equal parts.

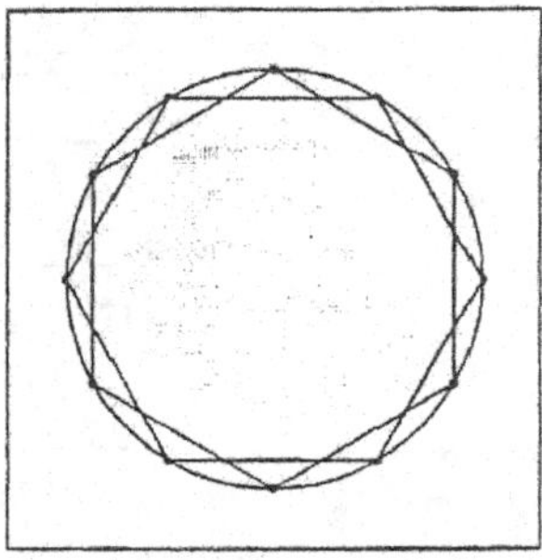

6. Connect every second point to produce two overlapping hexagons.

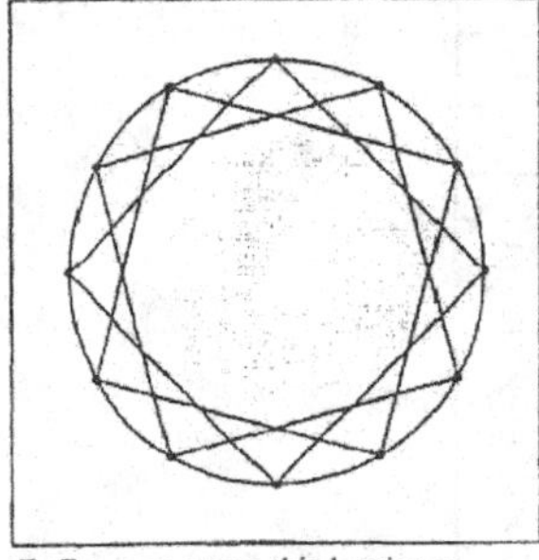

7. Connect every third point to produce three overlapping squares.

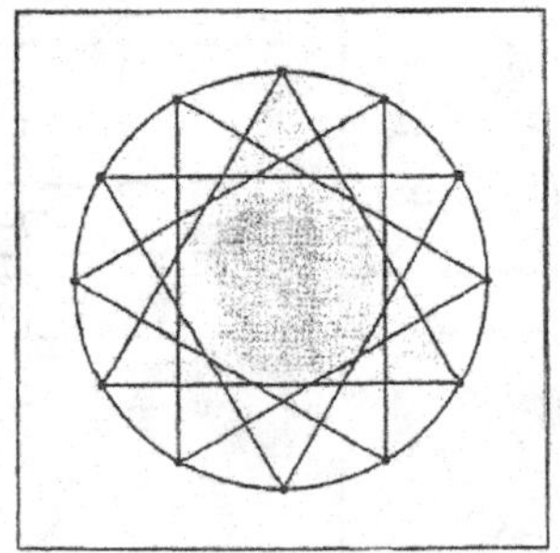

8. Connect every fourth point to produce four overlapping triangles.

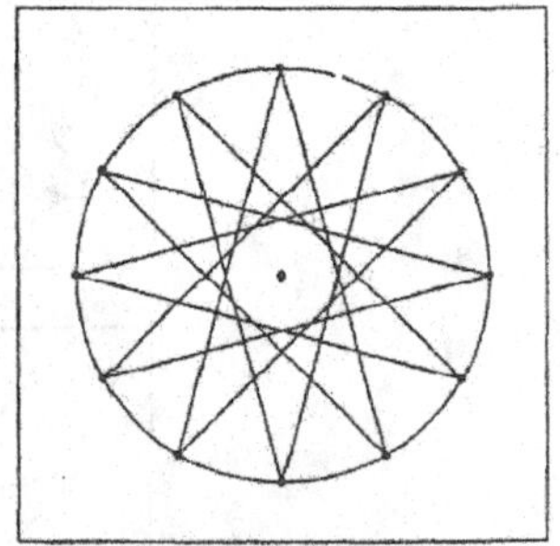

9. Connect every fifth point without lifting the pencil from the paper to produce a "true stellation."

Fig. 1.9. Drawing 6- and 12-pointed stars.

Construction 1: Use Adobe Illustrator, Corel Draw or simply a compass and straightedge to create a triangle-circle grid and a triangle grid beginning with a circle. After doing this, create your own pattern based on the triangle-circle grid and the triangle grid.

Stars

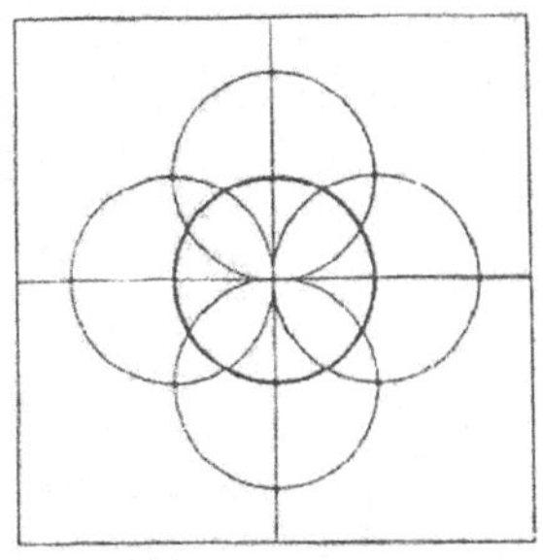

1. Eight-pointed and sixteen-pointed stars are related to the square grid construction. (See previous panel.) Connect the center...

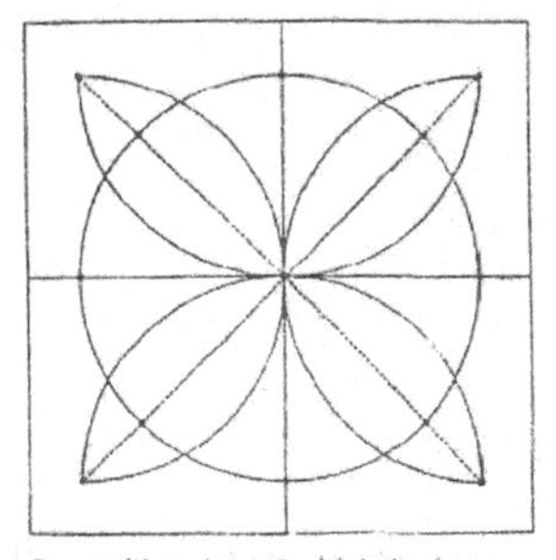

2. ...with points at which the four outer circles intersect.

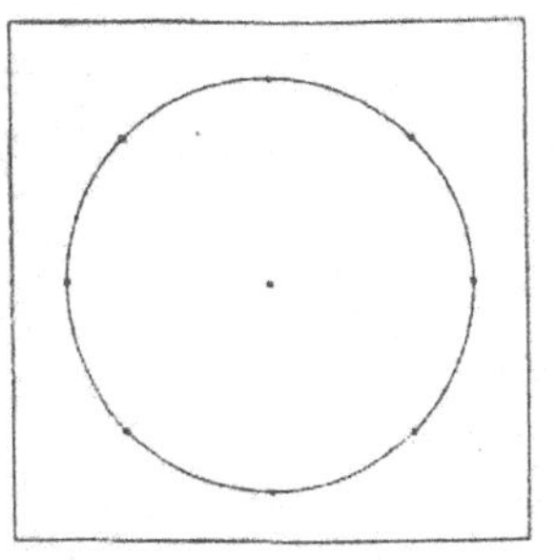

3. This divides the circle in eight equal parts.

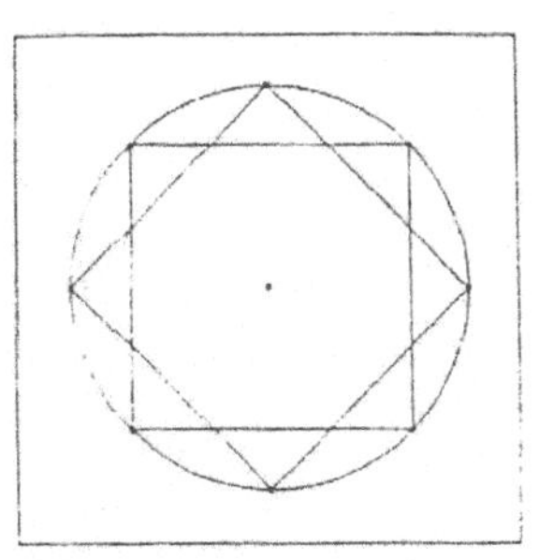

4. Connect every second point to produce an eight-pointed star with overlapping squares and an octagon in the center.

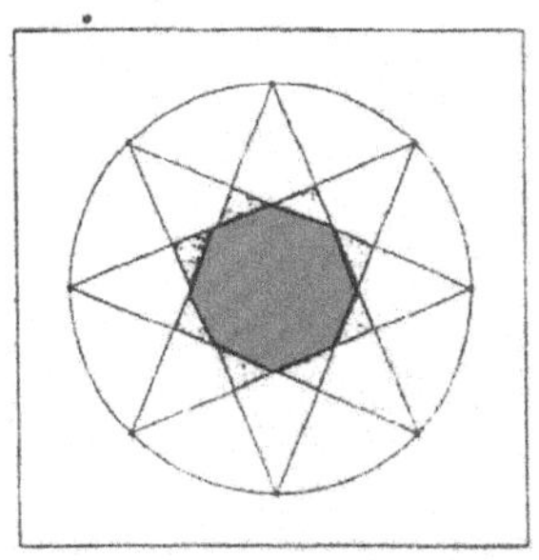

5. Connect every third point without lifting pencil from paper to create a "true stellation." Note that crossing lines produce the 8-pointed star of Fig. 4.

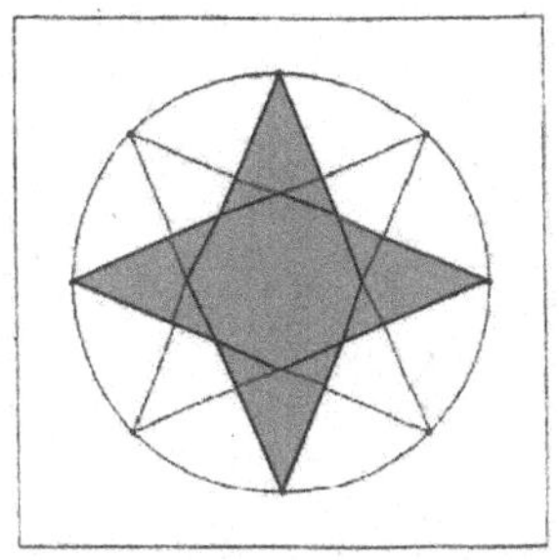

6. A popular four-pointed star is embedded in the eight-pointed star.

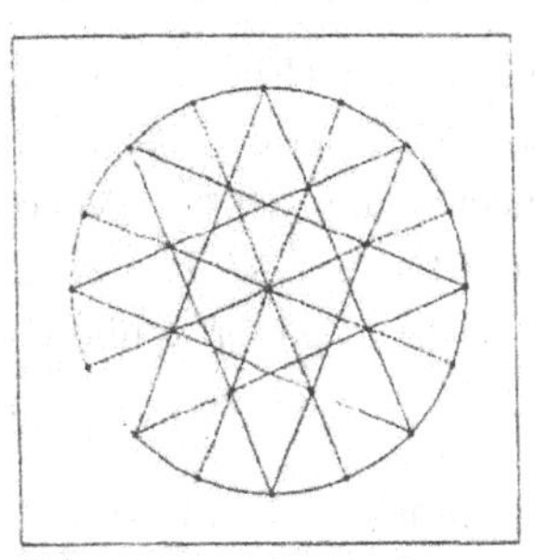

7. Connect opposite points in the central eight-pointed star to divide circle into sixteen equal parts.

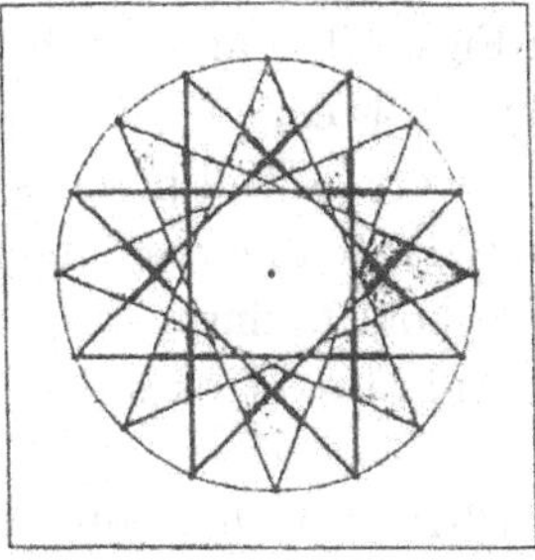

8. Connect every sixth point to produce a sixteen-pointed star with overlapping eight-pointed stars.

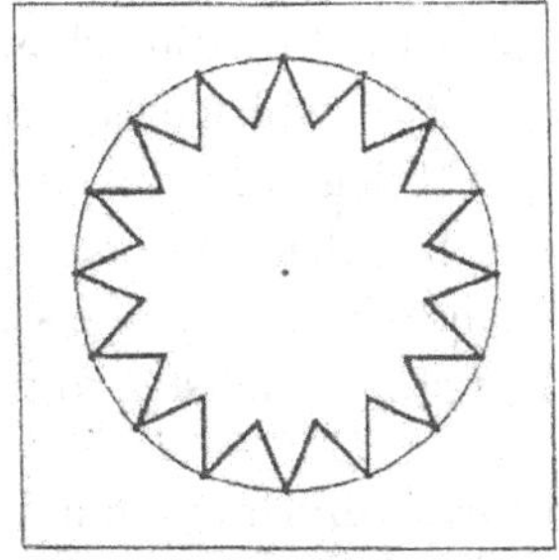

9. Stars were the favorite design element of Islamic artists.

Fig. 1.10. Drawing 8- and 16-pointed stars.

Construction 2: Beginning with a circle, repeat the instructions for Construction 1 to construct a pattern of your own based on the square-circle and square grids.

Construction 3: Repeat the instructions for Construction 1 and create star patterns based on six-, eight-, twelve- and sixteen-pointed stars. Alternatively you can experiment with other stars of your choice. Be bold and try to create an outrageous star.

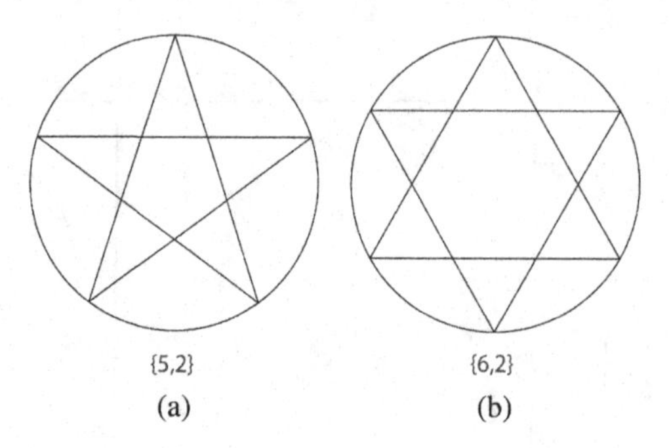

Fig. 1.11. (a) a {5,2} star; (b) a {6,2} star.

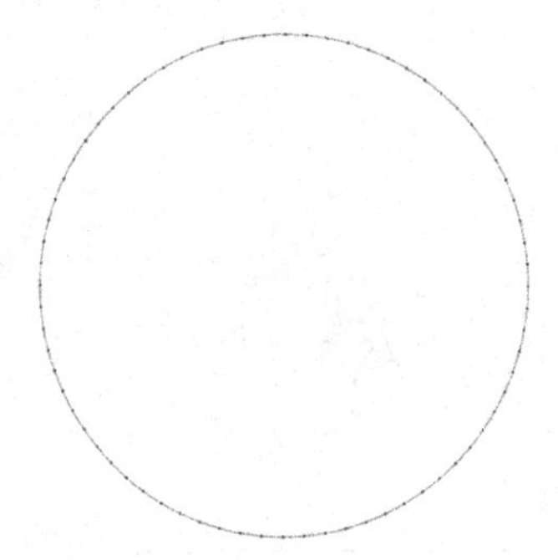

Fig. 1.12. A circle divided into 360 parts.

1.5. Star Exploration

Consider the two stars shown in Figs. 1.11a and 1.11b. The first star (Fig. 1.11a) has five points in which every second point is connected. I call this a $\{5,2\}$ star. The star in Fig. 1.11b has six points in which every second point is connected, i.e., $\{6,2\}$. Do you notice that there is something strikingly different about these two stars? The $\{5,2\}$ star can be drawn in one stroke without taking your pencil off of the paper, while the $\{6,2\}$ requires the star to be drawn in two strokes.

Exploration: Draw a number of $\{n,p\}$ stars, i.e., stars with n-points evenly distributed around a circle in which every p-th point is connected. Try to come up with a condition on n and p that enables you to predict whether an $\{n,p\}$ star can be drawn in a single stroke and how many times you have to take your pencil off of the paper to draw the star. You can use the circle with 360 divisions on it in Fig. 1.12 to help you in your exploration. Enlarge this star and replicate it.

1.6. The Two Great Systems of Ancient Geometry

The triangle-circle and square-circle patterns can be thought of as coming from two universes of ancient design both constructible with a compass and a straightedge.

a. Triangle-circle patterns lead to equilateral triangles. Dropping an altitude from a vertex to the base of such a triangle leads to a 30,60,90-deg. triangle.
b. Square-circle patterns lead to squares. Dividing a square by its diagonal leads to a 45,45,90-deg. triangle.

These two systems were of great importance in the construction of the great architectural masterpieces of antiquity. They will play a large role in the remainder of this book.

Five- and ten-pointed stars are also of great importance and will be introduced in Chap. 17.

CHAPTER 2

THE PYTHAGOREAN THEOREM

2.1. Introduction

The Pythagorean Theorem is perhaps the most important theorem in all of Euclidean geometry. No one knows its origin; however, it was probably known long before the age of Pythagoras (570–495 BC). There have been more than seven hundred proofs of this great theorem. This chapter describes five of them.

The Pythagorean Theorem basically states that if you place a square on each of the sides of a right triangle, the area of the square on the hypotenuse equals the sum of the areas of two squares on the other sides of the triangle (see Fig. 2.1a). For that matter, as Fig. 2.1b shows that you can just as well place Pythagoras on each edge of the triangle, the area of the Pythagoras on the hypotenuse then equals the sum of the areas of the Pythagorases on the other two sides.

2.2. Five Proofs of the Pythagorean Theorem

2.2.1 Pythagorean Theorem by dissection of a square

In Fig. 2.2a, you see a right triangle with area A_T and sides a, b, c. In Fig. 2.2b, an outer large square with area A_L is divided into 5 areas, four triangles of total area $4A_T$ and a smaller interior square of area A_S.

a. Express the following areas in terms of a, b and c:

$$A_S = \_\_\_\_\_, \quad A_T = \_\_\_\_\_, \quad A_L = \_\_\_\_\_.$$

b. Replace the expressions from part (a) into the equation:

$$A_L = A_S + 4A_T$$

which states that the sum of the five areas equals the area of the outer square.

c. After some algebraic manipulation, this should result in a proof of the Pythagorean Theorem for the original right triangle.

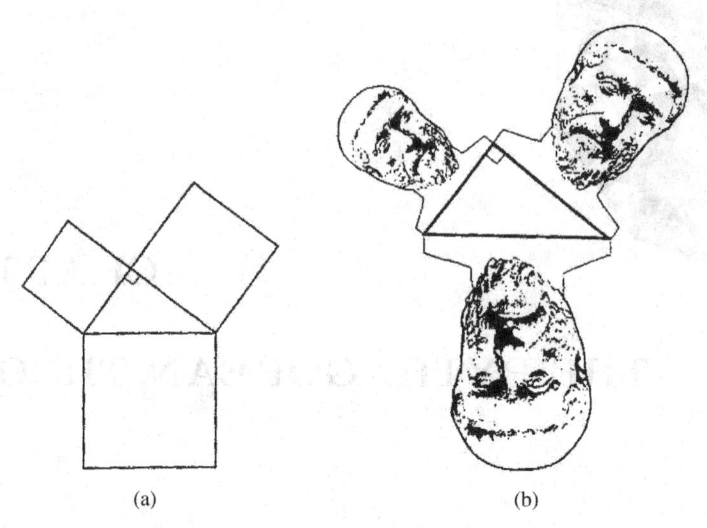

Fig. 2.1. (a) The area of the square on the hypotenuse of a right triangle equals the sum of the areas on the other two sides; (b) The same holds for Pythagoras.

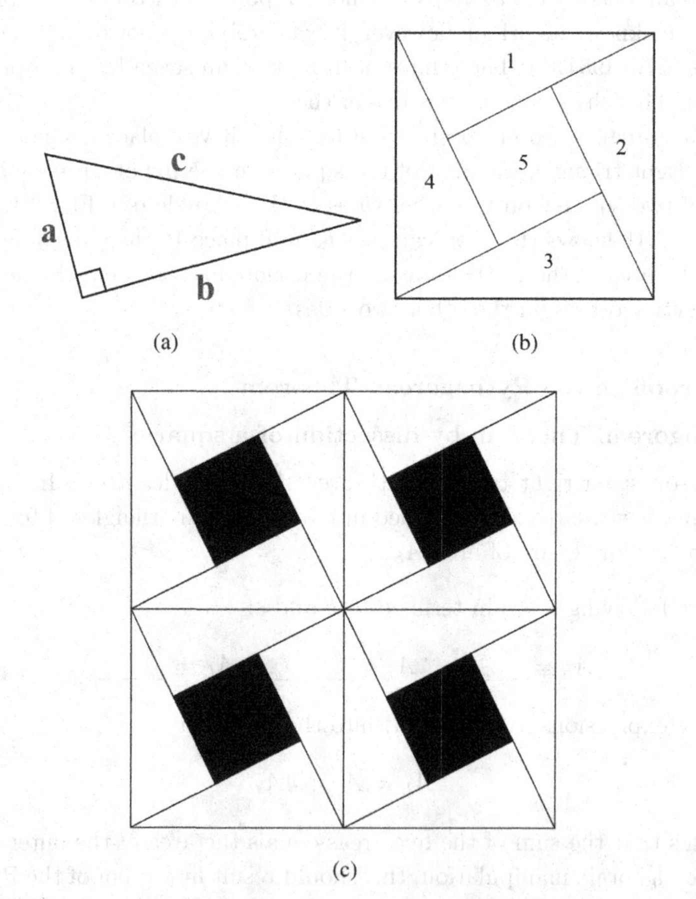

Fig. 2.2. (a) A right triangle labeled a,b,c ; (b) The area of a large square equal to the sum of the areas of a small square and four right triangles; (c) A design suggested by the quilter, Elaine Ellison.

d. In Fig. 2.2c, four of these squares are combined. Create a design by coloring the areas in this figure. This should work for any right triangle so that you can recreate this by cutting sixteen identical right triangles from a piece of construction paper.

2.2.2 Pythagorean cake cuts

In Fig. 2.3, a diagram has been drawn to represent a cake with four congruent slices [Gar]. The edge lengths of each cake slice are labeled a, b and c.

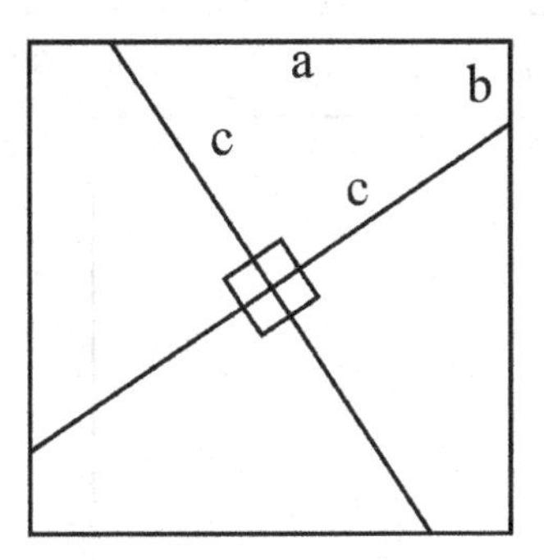

Fig. 2.3. Four identical cake slices with edge lengths a,b,c.

a. Reproduce Fig. 2.3 and use it to cut out the four slices. Before cutting, label each edge a, b, c.
b. Try your hand at reconstructing the original square from the four slices. Hint: The right angles should be placed at the corners of the square.
c. Now rearrange the cake slices to form two squares. Hint: Note that each slice has two right angles. The other set of four right angles should be placed at the corners of a large square. You should now have a large outer square, the four cake slices and an empty inner square.
d. Next express in terms of a, b and c, the areas of the outer square A_L, the combined cake slices which equals to the area of the original square cake A_C, and the small inner square A_S:

$$A_L = \_\_\_\_\_, \quad A_S = \_\_\_\_\_, \quad A_C = \_\_\_\_\_.$$

e. Replace the expressions in part (d) into the equation:

$$A_L = A_S + A_C$$

which expresses the fact that the outer area is the sum of the small empty area and the area of the original cake slices.
f. After some algebraic manipulation, you should be able to derive an expression which looks like the Pythagorean Theorem. Your challenge is to find the triangle within the original cake slices to which this expression pertains.

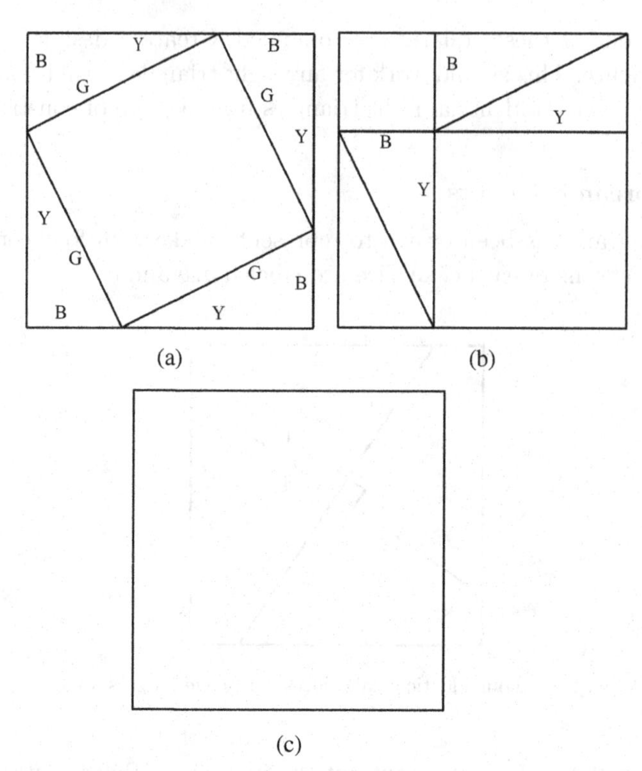

Fig. 2.4. (a), (b) An AHA proof of the Pythagorean theorem; (c) Cut out the four triangles and use the blank square to illustrate the theorem.

2.2.3 An AHA proof

This proof of the Pythagorean Theorem is an AHA proof. This means that the validity of the theorem should be evident by just looking at the diagram. In Fig. 2.4a, four congruent right triangles have been arranged in the outer square, leaving an inner square area empty.

a. First, take a magic marker and color code each edge of the right triangle, e.g., color the small side blue B, the long side yellow Y, and the hypotenuse green G.
b. In Fig. 2.4b, the triangles have been rearranged, leaving two square areas empty. You can carry this out by reproducing Fig. 2.4a and cutting out the color-coded triangles and arranging them into the empty square in Fig. 2.4c so that it looks like the configuration in Fig. 2.4a.
c. Now, rearrange the triangles to the positions shown in Fig. 2.4b.
d. The proof of the Pythagorean Theorem should now be evident. Explain why.

$$G^2 = B^2 + Y^2$$

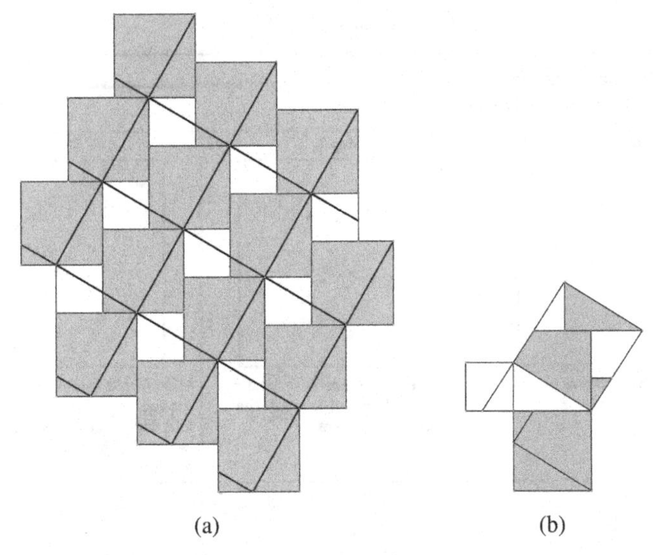

Fig. 2.5. (a) An Islamic design with squares of three different sizes; (b) Detail of how the large square can be dissected to form the other two squares.

2.2.4 Pythagorean Theorem by way of an Islamic tiling

Here is another AHA proof. Figure 2.5a shows a design attributed to Annairizi of Arabia (circa 900) tiled by a number of white and gray squares. Notice that the design is also tiled by large squares containing dissections of the white and gray squares. In Fig. 2.5b, these three sizes of squares are shown to be squares on the sides of a right triangle with the large square being the square on the hypotenuse. By observing this design, it is evident that the area of the square on the hypotenuse is the sum of the areas of two smaller squares, and therefore the Pythagorean Theorem is again proven. Do you see this? This design could have been made with any right triangle and the three squares on its sides. Figure 2.5b shows how the large square on the hypotenuse can be dissected to create the pieces that form the two smaller squares. Do you see this?

a. Choose any right triangle of your own.
b. Recreate Fig. 2.5b from your right triangle.
c. Cut out the polygons that make up the area of the large square and show that they recreate the two smaller squares by glueing them together to form these two squares.
d. Take several copies of the large square with its subdivisions and juxtapose them side by side to form the original pattern from Fig. 2.5a.
e. Repeat this construction for the case of a 45,45,90-deg. triangle.

2.2.5 An origami proof of the Pythagorean Theorem

An origami approach to the Pythagorean Theorem is shown in Fig. 2.6. Follow the instructions and show how this paper folding exploration amounts to the proof of the Pythagorean Theorem given in Sec. 2.2.1.

Paper-Folding Explorations

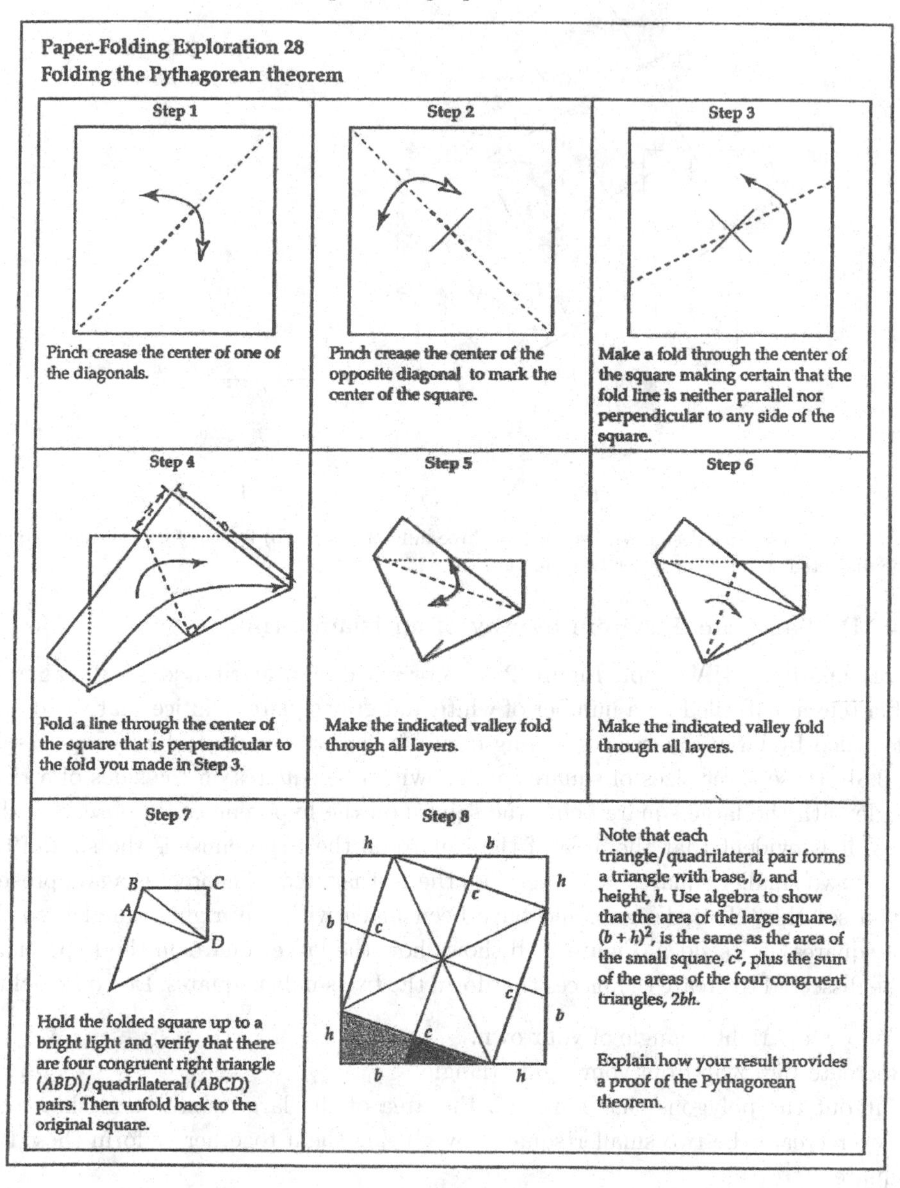

Paper-Folding Exploration 28
Folding the Pythagorean theorem
Step 1
Pinch crease the center of one of the diagonals.
Step 2
Pinch crease the center of the opposite diagonal to mark the center of the square.
Step 3
Make a fold through the center of the square making certain that the fold line is neither parallel nor perpendicular to any side of the square.
Step 4
Fold a line through the center of the square that is perpendicular to the fold you made in Step 3.
Step 5
Make the indicated valley fold through all layers.
Step 6
Make the indicated valley fold through all layers.
Step 7
B
C
A
D
Hold the folded square up to a bright light and verify that there are four congruent right triangle (ABD)/quadrilateral (ABCD) pairs. Then unfold back to the original square.
Step 8
h
b
h
b
c
c
c
c
b
h
b
h
Note that each triangle/quadrilateral pair forms a triangle with base, b, and height, h. Use algebra to show that the area of the large square, (b + h)², is the same as the area of the small square, c², plus the sum of the areas of the four congruent triangles, 2bh.
Explain how your result provides a proof of the Pythagorean theorem.

Fig. 2.6. An origami approach to recreating the proof in Sec. 2.2.1.

2.3. Pythagorean Triples and the Brunes Star

Certain right triangles can be formed by three integers such as a 3,4,5-right triangle, a 5,12,13-right triangle or an 8,15,17-right triangle. Of all these right triangles, the 3,4,5-triangle has distinguished itself and was revered in Egyptian times where it was used as the proportions of the sarcophagus in which the Pharaohs were buried. 3,4,5-relationships in general lie behind the structure of the musical scale and the color wheel [Kap 2].

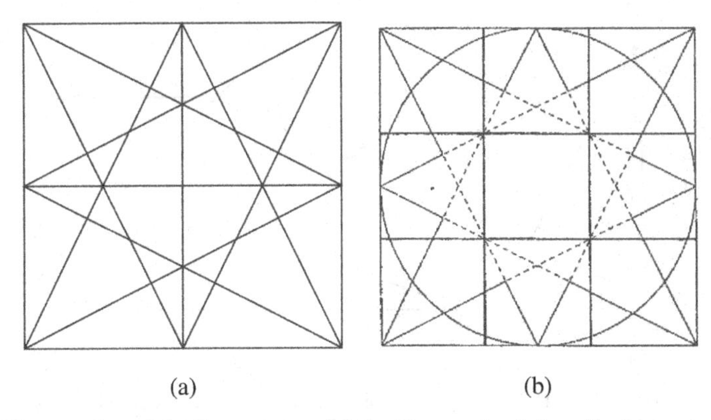

Fig. 2.7. (a) The creation of the Brunes star; (b) the Brunes star is inscribed in a circle and a square trisecting the base of the square and forming a nine square grid.

You can see 3,4,5-triangles emerge from a simple diagram (see Fig. 2.7a).

a. Take a square and divide it into two half-squares vertically. Draw the two diagonals in each half square.
b. Next divide it into two half-squares horizontally and place the diagonals into each of these half-squares. You now have a square tiled by many triangles. Observe that there are 3,4,5-triangles at four different scales.
c. Observe that this pattern can be formed from four large 3,4,5-triangles incident to each of the four vertices of the original square. Take a string and divide it into 12 equal parts. The circle can then be transformed to a 3,4,5-triangle. Therefore the Brunes star can be created from four loops of string.
d. If you place the two diagonals of the original square into this diagram, you can observe that at appropriate levels, the diagram divides the side of the square into $2, 3, 4, \ldots, 11, 12$ equal lengths. In Fig. 2.7b, the Brunes star is inscribed in a circle and a square and trisects the sides of the square dividing the square into a nine-square grid. Only the level of seven equal parts is ambiguous; it occurs at the level where the arc of a circle is drawn from the bottom left vertex of a square through the center of the square. In Chap. 17, we will see that this construction cuts the side of the square in what is called the *sacred cut.*
e. Enhance the power of this diagram by coloring it. Notice the eight-pointed star pattern within the square.

We refer to this construction as a *Brunes star* after the Danish engineer, Tons Brunes, who has shown that this configuration may have formed a template for many of the ancient temples of Greece and Rome [Bru], [Kap2].

2.4. Application of the Pythagorean Theorem

The Pythagorean Theorem can be used to find the distance AB across a lake from the position of two people standing at A and B as shown in Fig. 2.8a without having to get wet. A third person stands at position C where angle $\angle ACB$ is a right angle. Have

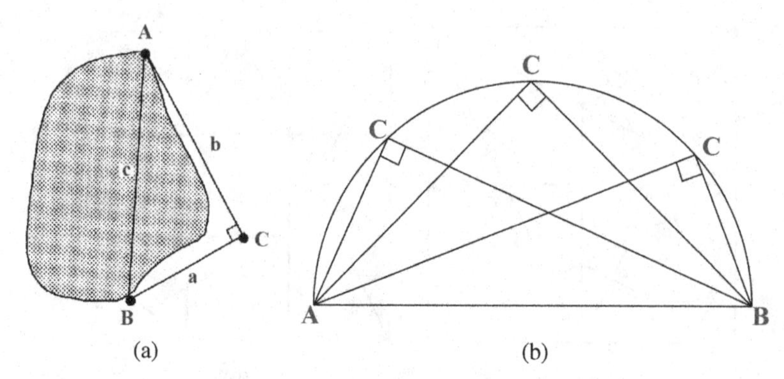

Fig. 2.8. (a) Application of the Pythagorean Theorem to finding inaccessible lengths, (b) application of the geometry of a circle to constructing a right angle.

someone walk off the distance from A to C and B to C on dry land. The distance from A to B can then be found using the Pythagorean Theorem. The one difficulty here is to find the exact position C for which $\angle ACB$ is a right angle. To help in this regard, you can use an angle mirror device in which two mirrors inclined to each other at 45 deg. are placed in a housing, erected on a flat platform atop a vertical pole. Looking into the mirrors, when persons A and B are simultaneously seen in the mirrors, that position must be a right angle, hence the location of C. Of course, there are many positions in which A and B can be simultaneously seen. We will prove in Chap. 8 that these positions lie on the circumference of the circle for which AB is the diameter as illustrated in Fig. 2.8(b). Why the mirrors should be inclined at 45 deg. will be discussed in Chap. 19.

CHAPTER 3

SCAN CONVERTING OF A LINE SEGMENT

3.1. Introduction

Mathematics considers a point to be of infinitesimal radius and a line to be of infinitesimal thickness. Point and line are therefore abstractions of the mind. In reality, what we generally consider to be a point has finite size and is rendered as a small dot or pixel while a line has finite thickness so that we can see it. Although pixels are used to represent points and lines, they do not satisfy the axioms of Euclidean geometry. To see this, one of the axioms of Euclidean geometry states that every line segment has a midpoint. Consider the line segment in Fig. 3.1a made up of five pixels. Clearly, the third pixel is the midpoint. However, the line segment in Fig. 3.1b, made up of six pixels, has no midpoint.

In this chapter, we will explore how the computer is able to create a line out of a set of pixels [Mey]. If a line extends from a pixel at the starting point $S(s_1, s_2)$ to a pixel at the endpoint $E(e_1, e_2)$, which pixels along the way must light up to form the best approximation to the abstract mathematical line?

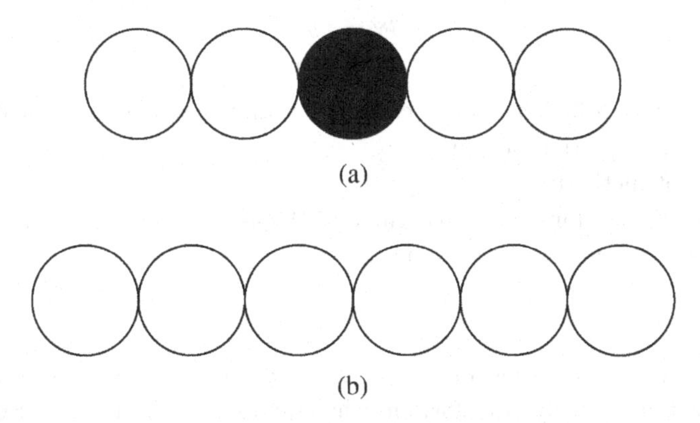

Fig. 3.1. A line made up of pixels (a) having a midpoint; (b) not having a midpoint.

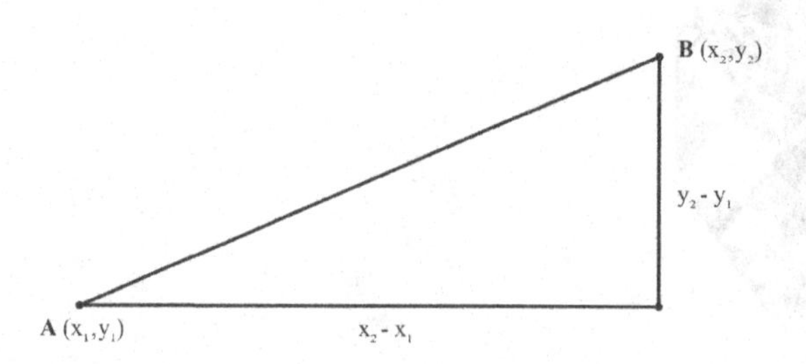

Fig. 3.2. The distance between points.

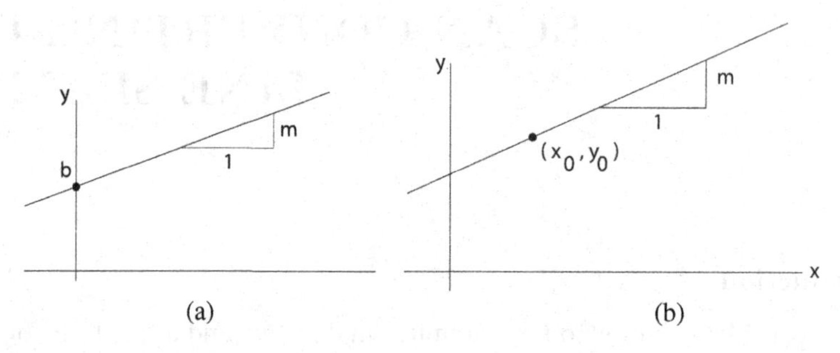

Fig. 3.3. Lines by (a) the slope, y-intercept method; (b) the point slope method.

3.2. Lines

First we review the mathematics of lines.

a. Consider a pair of points $(x_1, y_1), (x_2, y_2)$ defining line segment AB shown in Fig. 3.2. You will notice that they form a right triangle with base $(x_2 - x_1)$ and height $(y_2 - y_1)$. From the Pythagorean Theorem, the distance between the points or the length of line segment AB is,

$$D = \sqrt{(x_2 - x_1)^2 + (y_2 - y_1)^2}, \tag{3.1}$$

while the slope m of line segment AB is defined as,

$$m = \frac{rise}{run} = \frac{y_2 - y_1}{x_2 - x_1}. \tag{3.2}$$

In any equation of a line, x and y occur to the first power. I will now describe two ways to find the equation of a line.

b. Slope-Intercept method:
Here, we specify the slope, $m = \text{rise/run}$ and the y-intercept, b as shown in Fig. 3.3a in which case the equation of the line is,

$$y = mx + b. \tag{3.3}$$

If the line is given in another form, e.g., $2x + 3y = 6$, the slope and y-intersect can be found by algebraically transforming the equation to the form of Eq. (3.3), i.e., $y = -\frac{2}{3}x + 2$ in which case $m = -2/3$ and $b = 2$.

c. Point-Slope method:
Here we specify one point on the line (x_0, y_0) and slope m, as shown in Fig. 3.3b, in which case the line is given by,

$$y - y_0 = m(x - x_0). \tag{3.4}$$

The point-slope formula is generally the more useful of the two formulas for a line. Sometimes two points $(x_1, y_1), (x_2, y_2)$ on the line are given instead of one point and the slope in which case the slope can be computed from Eq. (3.2), and either of the two points can serve as the given point in the point-slope formula.

d. Two lines, l_1 and l_2, are perpendicular if their slopes are negative reciprocals of each other, i.e., $m_2 = -\frac{1}{m_1}$.

Problems:

1. Find the point-slope and slope-intersect forms of the line given a point $(2, -1)$ and the slope $m = -2$.
2. Find the point-slope and slope-intercept forms of the line given a point $(2, -1)$ and $m = \frac{1}{2}$.
3. Draw the graphs of the lines in Problems 1 and 2 and show that they are perpendicular.
4. Find the slope and length of the line segment defined by the two points $(2, -1)$ and $(3, 4)$.
5. Find the equation of the line defined by the points $(2, -1)$ and $(3, 4)$.

3.3. Lines and Calculus

What makes calculus work is that any "smooth" curve, however complex, can be approximated locally in the vicinity of a point by a line tangent to the curve with the approximation getting better as you get closer to the point. This also works for curves that are mostly smooth (they have only a finite number of kinks). For example, Fig. 3.4a is a smooth curve. Notice that the nearer you get to any point on it, the more the curve looks like a line. Calculus reduces the study of curves, a complex undertaking, to the study of lines, one of the simplest mathematical entities.

The graph of a curve is shown in Fig. 3.4b that is smooth everywhere but at two points. In the neighborhood of each of the other points, the curve looks locally like a line. Calculus deals only with mostly smooth curves. However, there are very important curves which are nowhere smooth. These curves are called *fractals*. Fractals are better

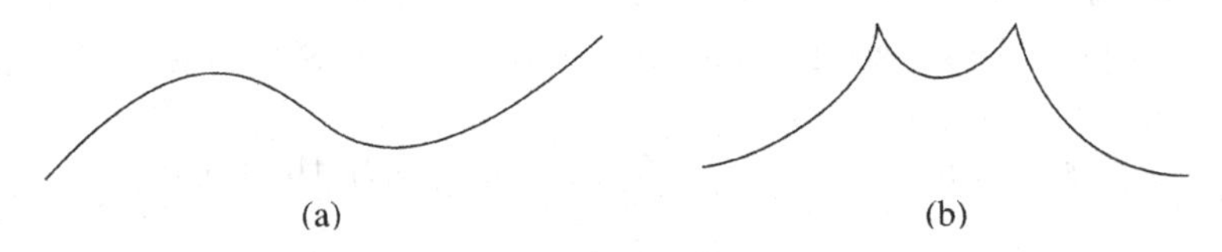

Fig. 3.4. Curves that are (a) smooth; (b) mostly smooth.

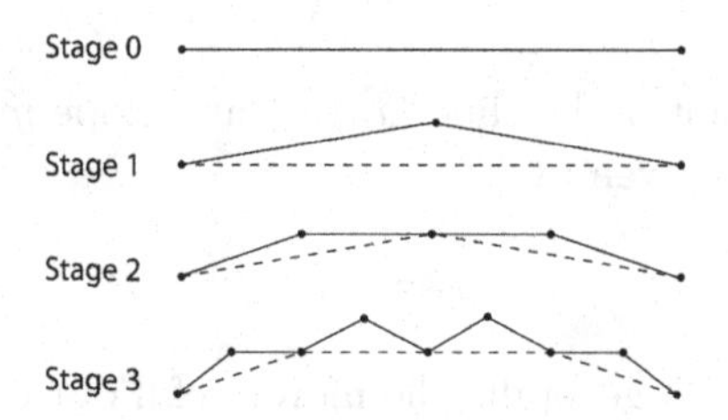

Fig. 3.5. Steps to the creation of a curve that is nowhere smooth.

in describing forms in nature such as the shape of coastlines, cloud formations and lightening bolts. Chapter 23 is devoted to an introduction to fractals.

The first three stages of a simple fractal are shown in Fig. 3.5. Begin with a straight line in Stage 0. In Stage 1, place a triangular roof on top of the line. In Stages 2 and 3, wherever there is a straight line segment, place a triangular roof atop with the same ratio of height to base. Try your hand at drawing several more stages to this fractal. Continue as far as your eye is able to discriminate line segments. There is no need to redraw the curve for each stage. You can just add each successive stage to your original line. If you were able to continue this ad infinitum, you would get a fractal curve that is nowhere smooth.

3.4. Scan Converting of a Line

A modern computer high definition screen is made up of a grid of 1200×1200 pixels of points in the first quadrant of a Cartesian coordinate system where (j, k) is a grid point for integer values $j, k \geq 0$. A pixel is then represented by a small circle drawn around each grid point as shown. Let us now consider the mathematical line drawn between the coordinates of the starting point of the line, $S(s_1, s_2)$ and the coordinates of the endpoint, $E(e_1, e_2)$. Using the point-slope formula,

$$y - s_2 = m(x - s_1) \tag{3.5}$$

where,

$$m = \frac{e_2 - s_2}{e_1 - s_1}. \tag{3.6}$$

This geometry of pixels is illustrated in Fig. 3.6 for the case of a line drawn between $S(2,1)$ and $E(7,3)$.

3.5. An Example of Scan Converting

Reconsider the line in Fig. 3.6 between the starting point $S(2,1)$ and the endpoint $E(7,3)$.

Since $s_1 = 2, s_2 = 1$ and $e_1 = 7, e_2 = 3$, using Eq. (3.2), the slope is,

$$m = \frac{3-1}{7-2} = \frac{2}{5} < 1.$$

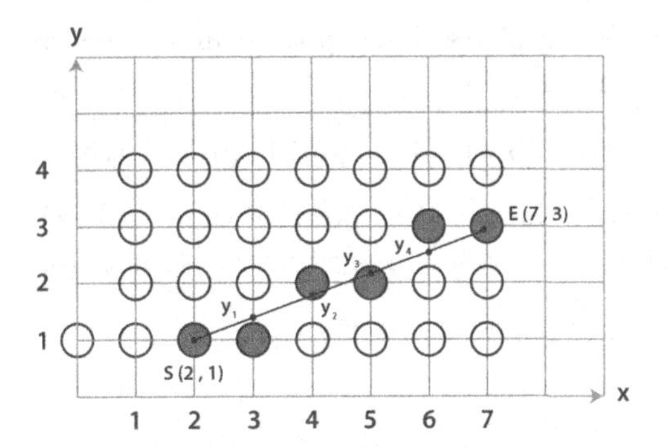

Fig. 3.6. A line approximated by pixels.

Replacing this in Eq. (3.5), the equation of the line is,

$$y - 1 = \frac{2}{5}(x - 2). \tag{3.7}$$

The x-coordinates of the points between S and E of this line are $x_1 = s_1 + 1, x_2 = s_1 + 2, x_3 = s_1 + 3, x_4 = s_1 + 4$ or, since $s_1 = 2$ it follows that $x_1 = 3, x_2 = 4, x_3 = 5, x_4 = 6$ while the corresponding y-coordinates are y_1, y_2, y_3, y_4, gotten by replacing the x values in Eq. (3.7). The values of x_k are integers but, in general, the values of y_k will not be integers.

Since the pixels all have integer values of x and y we have to convert the values of y_k to integers. To assign an integer value to the y-coordinate, we make use of the following *vertical-grid-line-crossing algorithm* which works so long as the slope m is less than or equal to 1 ($m \leq 1$) in absolute value:

Vertical-grid-line-crossing algorithm: Let y_k^* be the nearest integer greater than or less than its value y_k where the integer value of y is denoted by a star ($*$) as shown in Fig. 3.6, e.g., if $y_k = 2.7$ then $y_k^* = 3$, while if $y_k = 4.3$ then $y_k^* = 4$. If $y_k = 3.5$, then you can let y_k^* be either 3 or 4 but remain consistent for all such instances.

Replacing successive integer values of x in Eq. (3.7), the corresponding values of y are computed as follows:

$$y_1 = 1 + \frac{2}{5}(3 - 2) = 1.4,$$

and using the vertical-grid-line-crossing algorithm, $y_1^* = 1$

$$y_2 = 1 + \frac{2}{5}(4 - 2) = 1.8, \quad y_2^* = 2$$

$$y_3 = 1 + \frac{2}{5}(5 - 2) = 2.2, \quad y_3^* = 2$$

$$y_4 = 1 + \frac{2}{5}(6 - 2) = 2.6, \quad y_4^* = 3.$$

As a result the following pixels light up to form the line from the start pixel to the end pixel as shown in Fig. 3.6:

$$(2,1),\ (3,1),\ (4,2),\ (5,2),\ (6,3),\ (7,3).$$

What if the slope is greater than 1? Why does the above procedure break down? Can you modify this algorithm so that it works when the slope is greater than 1? But first solve the following problems.

Problems:

6. In a figure such as Fig. 3.6, given $S(5,12)$ and $E(10,14)$, find the integer coordinates of the pixels. You will have to determine the values of: $y_1^*, y_2^*, y_3^*, y_4^*$ as was done for the example in Sec. 3.5. Show on a graph which pixels light up.
7. In a figure such as Fig. 3.6, given $S(6,20)$ and $E(10,18)$, find the integer coordinates of all pixels.
8. As you can see in Fig. 3.6, the centers of the chosen pixels (shaded) do not necessarily lie on a straight line. In a figure of this sort, there are precisely three slopes for which the centers lie on a straight line when we carry out the vertical-grid-line-crossing algorithm. Determine what these slopes are and draw pictures of the lines to illustrate.
9. Draw a figure to illustrate why the vertical-grid-line-crossing algorithm presented above is not a good one for lines when the slope $m > 1$. Hint: Try to apply the algorithm to the line between $S(2,1)$ and $E(3,6)$.
10. Devise a variant of the vertical-grid-line-crossing algorithm that performs well if $m > 1$ and apply it to the values of S and E in Problem 9. Hint: Solve Eq. (3.5) for x. Let the values of y take on integer values and find the locations where x crosses horizontal lines.

11a. Given the center of a pixel, say (s_1, s_2), write an expression in terms of s_1 and s_2 for the centers of the four pixels that are neighbors of (s_1, s_2).

11b. Here is another scan-conversion algorithm. First find the centers of the four neighbors of the start pixel. If any of them is E, no intermediate pixels should light up. Otherwise find a neighbor where the distance from its center point to E is the shortest and light up that pixel. In case of a tie, pick one at random. Continue this process until you reach the end pixel. Carry this out for $S(3,2)$ and $E(6,3)$. As you do this, keep track of the number of additions, subtractions and multiplications that you require. Compare it to the number of arithmetical operations required by the vertical-grid-line-crossing algorithm. The fewer the number of operations the more efficient the algorithm.

CHAPTER 4

COMPASS AND STRAIGHTEDGE CONSTRUCTIONS

PART 1: THE WORLD WITHIN A TRIANGLE

4.1. Geometric Constructions

Compass and straightedge constructions have always been a favorite topic in geometry. With straightedge and compass alone, a great variety of constructions can be performed as the reader may remember from high school: a line segment or angle can be bisected; a line can be drawn from a point perpendicular or parallel to a given line, a regular triangle, pentagon, hexagon, octagon, decagon and dodecagon (12 sides) may be inscribed in a circle; and circles can either be inscribed in or circumscribed about a triangle. In all these problems, the ruler is used merely as a straightedge: an instrument for drawing a straight line, but not for measuring or marking off distances. The traditional restriction to straightedge and compass goes back to antiquity, although the Greeks did not hesitate to use other instruments. There is strong evidence that the great architecture of the ancient world was carried out with a length of rope functioning as a compass and a rod acting as a straightedge. Grout lines have been found on the floors of temples and cathedrals marked by the rope compass.

It was a great puzzle to ancient mathematicians that, although regular polygons with $n = 3, 4, 5, 6, 8, 10$ and 12 sides could be constructed with compass and straightedge alone, a heptagon (7-sided polygon), a nonagon (9-sided polygon) and an 11-sided polygon could not be constructed despite great efforts. Also a given angle could not be trisected, or a cube whose volume was twice a given volume, constructed. After centuries of futile search for solutions to these problems, mathematicians were able to prove, by using algebraic methods, that they cannot be done. In this section, we will not go through these proofs which are nicely stated in What is Mathematics [Cou] by Courant and Robbins. Instead, in this chapter, we will show that even a geometric object as simple as a triangle has a world of relationships residing within it. We will present geometrical constructions with no explanation as to why they work. In Chaps. 7, 10 and 12 we will present geometrical proofs to justify these constructions.

You will need a compass and straightedge.

4.2. The Fundamental Constructions

Carry out the following three constructions:

4.2.1 Construction 1: Find the perpendicular bisector of a line segment AB

a. Place the compass point at A and draw an arc as shown in Fig. 4.1. Then draw the same length of arc from B, and find points C and D, the intersection of these arcs.
b. Connect C and D with a line. This line is the perpendicular bisector of AB.

4.2.2 Construction 2: Find the line through P perpendicular to a given line l

Method 1:

a. Draw the arc of a circle with center at P intersecting line, l at points A and B as shown in Fig. 4.2a.
b. Use the method in Construction 1 to find the perpendicular bisector of AB which goes through P.

Method 2:

a. Choose arbitrary points A and B on line l as shown in Fig. 4.2b. With centers at A and B, draw arcs of a circle through P.
b. The two arcs intersect at P and C. The line PC through P is perpendicular to l.

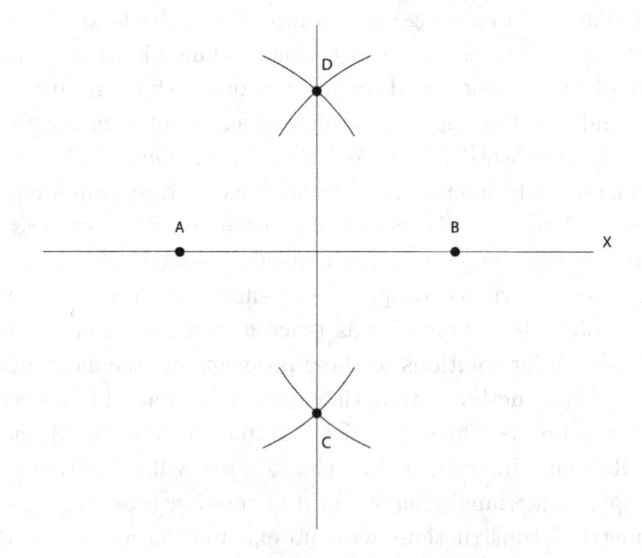

Fig. 4.1. Construction of an angle bisector.

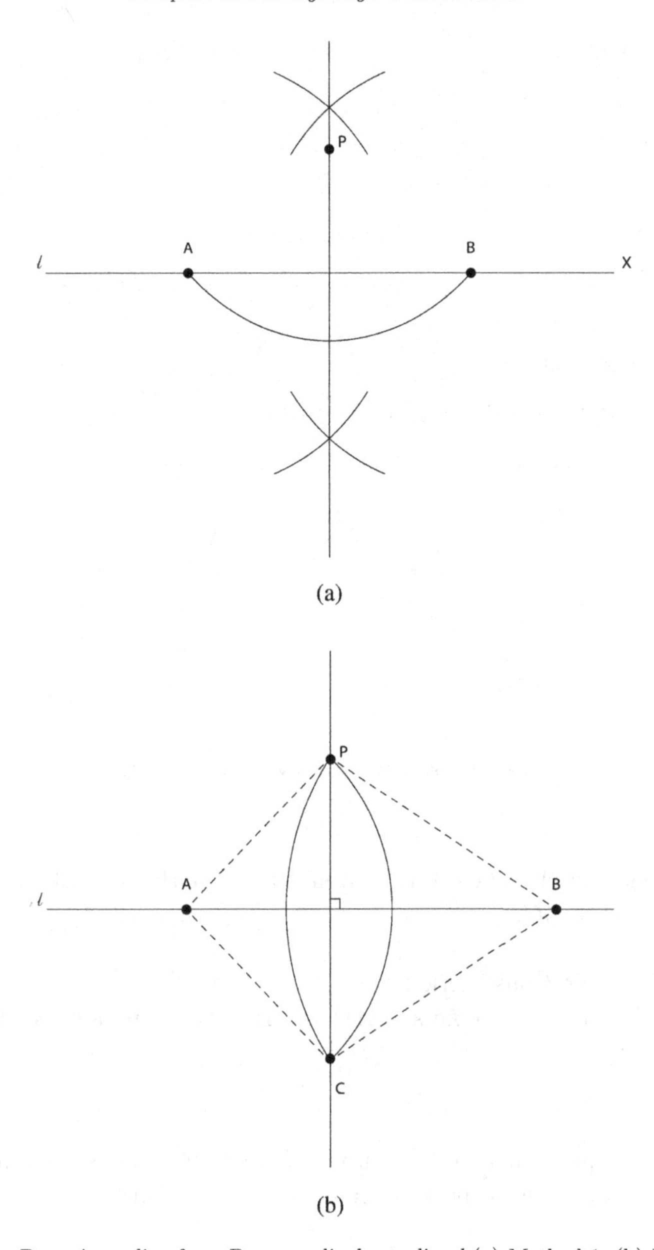

Fig. 4.2. Dropping a line from P perpendicular to line l (a) Method 1; (b) Method 2.

4.2.3 Construction 3: Draw a line through a given point C parallel to a given line l

Method 1:

a. Choose arbitrary points A and B on line l as shown in Fig. 4.3a.
b. Draw line AC.

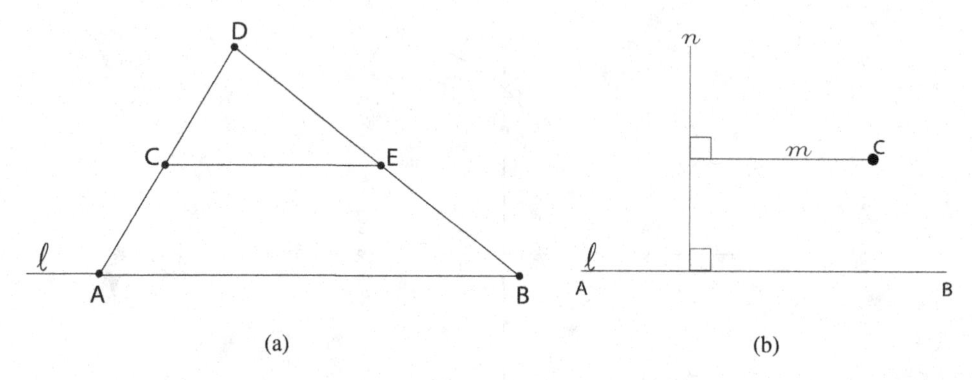

Fig. 4.3. Constructing a line through point C parallel to line l (a) Method 1; (b) Method 2.

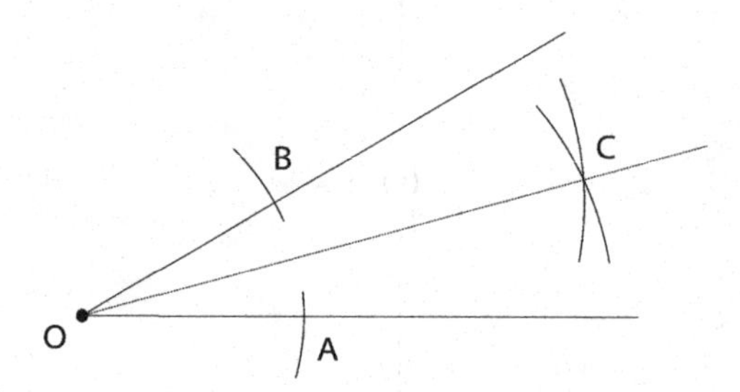

Fig. 4.4. Constructing the bisector of an angle.

c. Extend line segment AC and, on the extended line mark off D where segment CD is of length equal to AC.
d. Draw line BD.
e. Bisect line BD using Construction 1.
f. With C and E as bisection points of AD and BD, CE is parallel to AB.

Method 2:

a. Draw a line n perpendicular to line l using Construction 2 as shown in Fig. 4.3b.
b. From point C, draw line m perpendicular to n using Construction 2. This line is parallel to AB.

4.2.4 Construction 4: Bisect a given angle

a. Given an angle as shown in Fig. 4.4, place your compass point on the vertex O and mark off equal line segments OA and OB on the *rays* that make up the angle.
b. Place your compass points on A and B and sweep out a pair of arcs that intersect at C.
c. Line OC is the angle bisector.

4.3. The World Within a Triangle

A triangle is a very interesting mathematical object. Within it are several special points. We now investigate these special points for an arbitrary triangle using Constructions 1–4.

a. **Construction 5:** The *median* of a triangle is a line from a vertex that bisects the opposite side. Show that the three medians of a triangle meet at a common point and that this point is at the trisection point of the three medians. The three medians mark the balance point or *centroid* of the triangle and divide the triangle into six triangles of equal area.
b. **Construction 6:** The *three perpendicular bisectors* of a triangle meet at a common point and this point is the center of the circle that goes through the vertex points. This circle is called the *circumscribing circle* and its center is referred to as the *circumcenter.*
c. **Construction 7:** The three *angle bisectors* of a triangle meet at a common point, and this point is the center of the circle that is tangent to the three sides of the triangle, the *inscribed circle*, and its center is referred to as the *incenter.*
d. **Construction 8:** An *altitude* of a triangle is a line drawn through a vertex of the triangle perpendicular to the opposite side. The three altitudes of a triangle meet at a common point referred to as the orthocenter. Note: If the altitude falls outside of the triangle, extend the opposite side to meet the altitude.
e. **Construction 9:** The circumcenter, orthocenter and centroid lie on a straight line called the *Euler line.*
f. **Construction 10:** There is another mysterious circle that goes through nine points of the triangle called the *9-point circle.* You can look this up on the Internet.

For a general triangle (not isosceles or equilateral), construct the special points and the Euler line from Constructions 5–10.

4.4. Special Points in an Equilateral Triangle

Constructions 11:

a. Consider a length of 5 boxes on a piece of graph paper to be 1 unit and construct an equilateral triangle with edge measuring 2 units.
b. Show that all of the special points: centroid, circumcenter and orthocenter, collapse to a single point.
c. Draw the inscribed and circumscribed circles.
d. Notice in Fig. 4.5a that the line from a vertex to the midpoint of the opposite side divides the equilateral triangle into two 30,60,90-triangles. Such triangles are important; their edge lengths are shown in Fig. 4.5b.
e. Using the properties of the 30,60,90-triangle in Fig. 4.5, find the area of the equilateral triangle, its perimeter, and the radii of the circumscribed and inscribed circles.

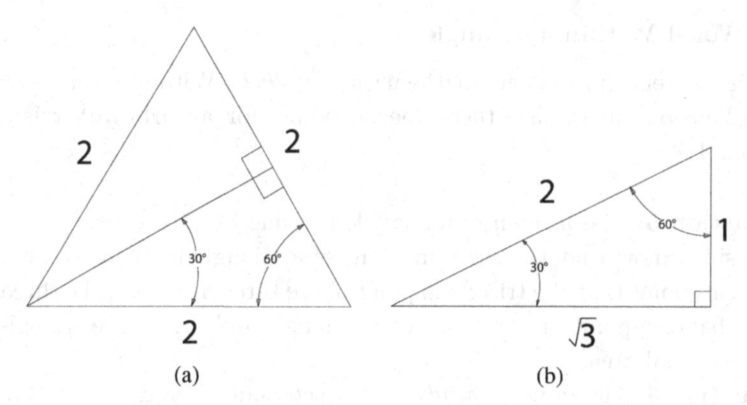

Fig. 4.5. (a) An equilateral triangle is bisected into two 30,60,90-triangles; (b) detail of a 30,60,90-triangle.

4.5. A Theorem about Inscribed Circles Within a Triangle

There is a wonderful theorem of Euclidean geometry that pertains to the inscribed circle of a triangle which states:

Theorem: *The product of the radius of the inscribed circle and the semi-perimeter of a triangle equals its area, i.e.,*

$$A = \frac{1}{2}pr$$

where p is the perimeter of the triangle and r is the radius of the inscribed circle.

Use this formula to find the radii of the inscribed circles of the 3,4,5-; 5,12,13-; and 8,15,17-right triangles. What do you think of these results? It is interesting that in some of my research, I discovered that these triangles play important roles in the proportions of the great Greek temple, the Parthenon.

Construction 12:

a. Draw these three right triangles to scale with their inscribed circles.
b. By direct measurement, verify that the radii are as predicted from the above formula.
c. Find the radius of the inscribed circle for an equilateral triangle and compare your result with Construction 11e.

4.6. Inscribing Regular Polygons in Circle

4.6.1 Construction 13: Inscribe a hexagon within a unit circle

a. Choose an arbitrary point A on the circumference of a circle (see Fig. 4.6).
b. Draw radius OA from the center O of the circle and define the length of OA as one unit.
c. Mark off from point A six chords of length OA.

This results in a hexagon.

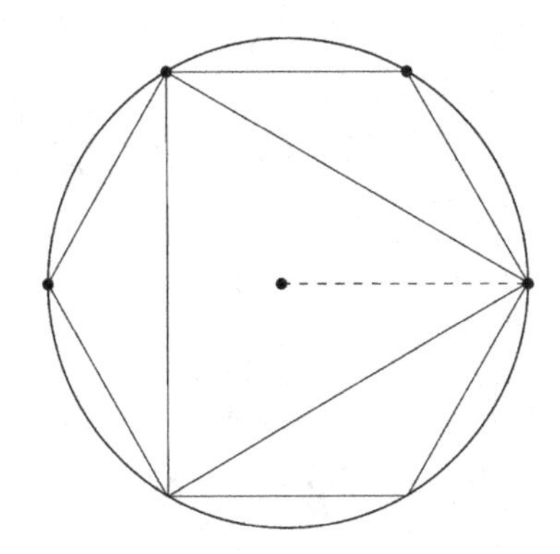

Fig. 4.6. A hexagon and an equilateral triangle inscribed in a circle.

4.6.2 Construction 14: Inscribe an equilateral triangle in a circle

To create an equilateral triangle inside a unit circle, construct a hexagon as shown in Sec. 4.6.1 and then connect every other vertex to form an equilateral triangle.

4.6.3 Construction 15: Inscribe a dodecagon inside a unit circle

a. Proceed as you did for a hexagon in Sec. 4.6.1.
b. Bisect each side of the hexagon.
c. Draw the radii through these points of bisection and mark the intersection points on the circle. Along with the original vertices of the hexagon, these new points on the circle form the vertices of the dodecagon.
d. Complete the dodecagon by connecting the vertices.

4.7. The Centroid of a Polygon

We first find the centroid of an irregular pentagon and then generalize to any polygon.

Construction 16: The centroid of an irregular pentagon.

Method 1:

a. Draw a pentagon on a piece of cardboard and use a pair of scissors to cut out your pentagon.
b. From an arbitrary point on the periphery, pin the pentagon to a wall.
c. Drop a plumbline from the pin and trace the plumbline on the face of the pentagon.
d. Pin another point on the periphery of the pentagon to the wall and drop another plumbline.
e. The centroid is where the two plumblines meet.

Method 2:

a. Label the vertices of the pentagon, A, B, C, D, E as shown in Fig. 4.7.
b. Bisect AB and let M_1 be the point of bisection.
c. Divide the line segment M_1C in three equal parts. Label the endpoint of the first segment M_2.

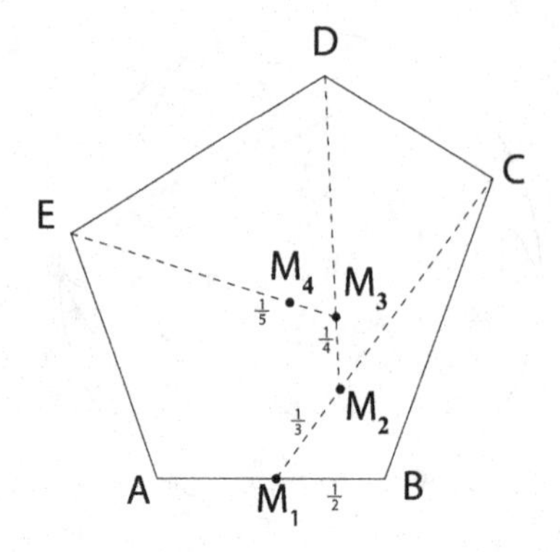

Fig. 4.7. Construction of the centroid of a pentagon.

d. Divide the line segment M_2D in four equal parts. Label the endpoint of the first segment M_3.
e. Divide the line segment M_3E in five equal parts. Label the endpoint of the first segment M_4.
f. M_4 is the centroid of the pentagon.

Find the centroid of the pentagon that you cut out in Method 1 of Construction 16 and check your result by using Method 2.

Remark 1: According to Construction 5, the centroid of a triangle can be found by first bisecting any side of the triangle and then trisecting the median to the opposite vertex so that the construction of the centroid of the pentagon in Method 2 of Construction 16 is seen to be a generalization of the centroid of a triangle.

Remark 2: An analogous construction to the one we carried out for a triangle and a pentagon can be carried out for polygons of any number of sides.

4.8. Additional Constructions

1. Construct an equilateral triangle given one of its sides (this yields the construction of a 60-deg. angle).
2. Given a point A on line l, construct a line through A perpendicular to l.
3. Bisect a given angle.
4. Double that given angle.
5. Construct a 30-deg. angle.
6. Given a line l and a point A not on l, construct a line m passing through A intersecting l at 30 deg.
7. Given one of the two diagonals of a square, construct the square.

CHAPTER 5

CONGRUENT TRIANGLES

5.1. Introduction

Two figures are said to be *congruent* if you can move one to the other without changing its shape so that either the first figure or its mirror image matches point for point with the second as shown in Fig. 5.1 for a triangle.

Consider a pair of triangles ΔABC and $\Delta A'B'C'$. We would like to know if these triangles are congruent. From the definition, they are congruent if their corresponding sides and angles are equal, i.e., $AB = A'B'$, $BC = B'C'$, $CA = C'A'$ and $\angle A = \angle A', \angle B = \angle B', \angle C = \angle C'$. We use the notation: $\Delta ABC \cong \Delta A'B'C'$ to denote congruency.

Note: We are using the convention that upper case letters represent angles while lower case letters refer to sides of triangles. The notation of a standard *reference triangle* is shown in Fig. 5.2 illustrating $\angle A, \angle B, \angle C$ and sides a, b, c opposite these angles.

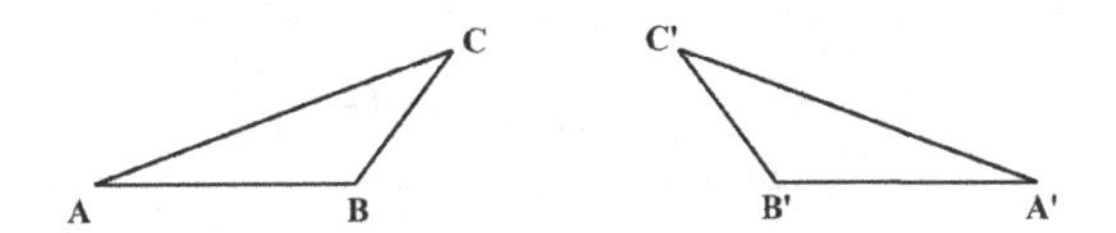

Fig. 5.1. Two congruent triangles.

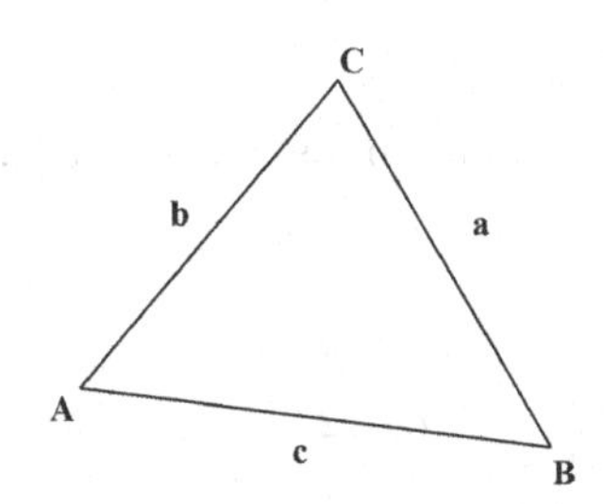

Fig. 5.2. A reference triangle naming sides a,b,c and angles A,B,C.

Notice that the sequence of sides and angles: $a, b, c, a, b, \ldots$ and $A, B, C, A, B, \ldots$ follow a counterclockwise order.

5.2. The Sum of the Angles of a Triangle

Draw a triangle other than isosceles or equilateral known as a *scalene triangle.* Cut off the three vertices, place their vertices together, and reassemble them to show that their angles sum up to 180 deg., a *straight angle.*

5.3. Conditions for Congruent Triangles

Given the following partial information about a triangle, can you create only a single triangle, more than one triangle, or no triangles from this information? On a piece of graph paper, use a box to represent 1 unit and construct the triangle if one is possible. It is helpful to refer to the reference triangle in Fig. 5.2. In what follows, by S we mean the length of a side of the triangle and by A we mean the measure of an angle.

5.3.1 Construction 1: SSS

a. $S_1 = 5$ units, $S_2 = 9$ units, $S_3 = 7$ units
b. $S_1 = 5$, $S_2 = 4$, $S_3 = 12$
c. $S_1 = 5$, $S_2 = 4$, $S_3 = 9$

The following information is used in the order given so that SAS is different from SSA. You will need a compass, straightedge, protractor, and graph paper.

5.3.2 Construction 2: SAS

By constructing a triangle with your own given information, show that SAS always results in a single triangle. Record the elements of your triangle below:

$$S_1 = c = \_\_\_\_\_\text{units}$$
$$\angle A = \_\_\_\_\_\text{deg.}$$
$$S_2 = b = \_\_\_\_\_\text{units}$$

5.3.3 Construction 3: SSA

Sometimes but not always, SSA results in a single triangle. As an example where SSA results in two triangles:

$$\angle A = 20\,\text{deg.}$$
$$c = 2$$
$$a = 1.$$

Construct two triangles with this partial information.

5.3.4 Construction 4: AAA

Show that AAA results in many triangles of the same shape but different sizes.

$$\angle A = \_\_\_\_\_ \text{deg.}$$
$$\angle B = \_\_\_\_\_ \text{deg.}$$
$$\angle C = \_\_\_\_\_ \text{deg.}$$

where $\angle A + \angle B + \angle C = 180\,\text{deg}$.

5.3.5 Construction 5: ASA

If the angles AAA are given along with one edge S, then there is only one triangle that can be drawn. Show that this can be done. This situation is usually referred to as AAS or ASA, since we have seen by the result of Sec. 5.2 that if two angles of a triangle are known, the third angle is also known because the three angles sum to 180 deg.

Construct a triangle given the partial information:

$$\angle A = 60\,\text{deg.}$$
$$c = 8 \text{ units}$$
$$\angle B = 45\,\text{deg.}$$

If only one triangle can be formed from the above partial information, then two triangles that agree in this partial information must be congruent. Therefore, from the above constructions, SAS, ASA and AAS insure that triangles agreeing in this partial information are congruent but SSA and AAA do not.

SSS also insures congruency provided that the *triangle inequality* holds.

5.4. Triangle Inequality

The sum of any two sides of a triangle must exceed the third side.

For example, Construction 5.3.1a above satisfies this inequality but Constructions 5.3.1b and 5.3.1c do not.

5.5. Constructing a Triangle from Partial Information

Let us now consider the following notation:

$$\angle A, \angle B, \angle C = \text{angle } A, \text{ angle } B, \text{ angle } C \text{ (see Fig. 5.2)}$$
$$a, b, c = \text{side opposite angle } A, \text{ side opposite angle } B, \text{ side opposite angle } C$$
$$h_a, h_b, h_c = \text{altitude to side } a, \text{ altitude to side } b, \text{ altitude to side } c$$
$$m_a, m_b, m_c = \text{median to side } a, \text{ median to side } b, \text{ median to side } c.$$

For the following problems, construct a triangle with the given information. Choose a number of boxes on your graph paper for a unit that results in a triangle not too large or small. Is the triangle unique or can you construct more than one? Determine the remaining angles and edge lengths by measurement with a ruler and protractor. I will demonstrate the solution to Problem 1 and ask you to solve for the sides and angles of the triangles derived from the partial information of the other problems. The results of some of the constructions will be given in Sec. 6.8.

In solving these problems, it will be helpful to refer to Fig. 5.2, the reference triangle.

Problems:

1. Let $\angle A = 45$ deg., c = 3, $h_c = 1$. Construct a unique triangle with the given information and determine the values of a,b, $\angle B$ and $\angle C$ by measurement with ruler and protractor.

Solution: Refer to the reference triangle shown in Fig. 5.2. Let side $c = 3$ be the line segment AB in Fig. 5.3. Since we are going to draw an altitude to AB from $\angle C$ it is best to orient AB on graph paper so that it is horizontal. Use a protractor or your graph paper to draw $\angle A = 45$ and draw the 45 deg. ray extending from vertex A. Construct a line segment DE of length $h_c = 1$ perpendicular to AB. Through E draw a line parallel to AB intersecting the ray of $\angle A$ (see Construction 2 from Sec. 4.2.3). This is the location of vertex C. Complete the triangle by drawing line segment BC. The other sides and angles can be found by measurement.

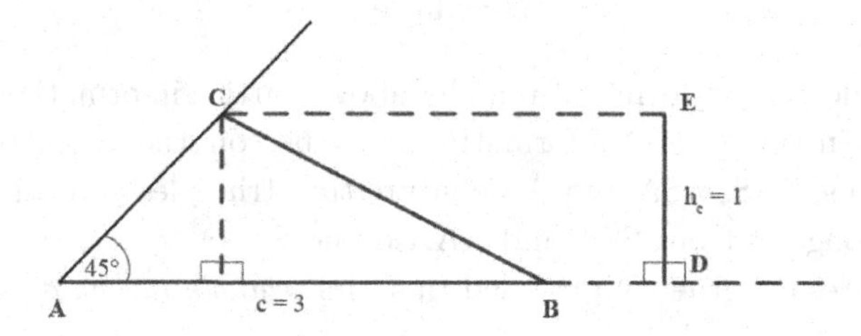

Fig. 5.3. An example of the construction of a triangle from partial information about it.

A word about measurement:

Let us say that 4 boxes on my graph paper correspond to 1 unit. Then $c = 3$ units or 12 boxes which on my ruler is 6 cm. After constructing the triangle, I measure side b to be 2.8 cm. To find b in units set up the following ratio,

$$b \text{ units}/b \text{ cm} = c \text{ units}/c \text{ cm} \quad \text{or} \quad \frac{b}{2.8} = \frac{3}{6}$$

Therefore $b = 1.4$ units. All measurements can be handled in a similar manner. We will be able to compute the exact value by trigonometry which is the subject of the next chapter.

2. Let $\angle A = 30$ deg., $c = 2$, $m_c = 1$. Determine a, b, $\angle B$ and $\angle C$
3. Let $a = 2$, $\angle A = 40$ deg., $h_c = 1$. Determine b, c, $\angle A$ and $\angle B$ and $\angle C$
4. Let $a = 4$, $b = 3$, $h_c = 2$. Determine c and all three angles.
5. Let $c = 4$, $m_c = 2$, $h_c = 1$. Determine a, b, and all three angles.
6. Let $\angle A = 50$ deg., $\angle B = 90$ deg., $b = 2$. Find c.
7. Let $\angle A = 40$ deg., $\angle B = 90$ deg., $a = 1$. Find b.

8. Let $\angle A = 150$ deg., $b = 1$, $c = 2$. Find a, $\angle B$.

9**. Let $\angle A = 60$ deg., $c = 2$, $a+b = 4$ construct a triangle with the given information. Can you do this given any values of c, $a + b$, and $\angle A$?

Note: Problem 9** is a big challenge. A solution to this problem will be presented in Chap. 7.

In the next chapter, you will be asked solve these triangles using trigonometry. You will find that whenever a triangle can be constructed with partial information, the length of the sides and angles can also be calculated by trigonometry.

5.6. Two Applications of Congruent Triangles

1. There is a boat moored offshore at point E. A person stands on dry land at point A where AE is perpendicular to the shoreline (see Fig. 5.4). The person would like to know the distance AE to the ship through an indirect measurement. The Greek philosopher Thales came up with a solution to this problem using congruent triangles. The person walked a distance of AB, perpendicular to AE and dug a spike into the ground. He then proceeded an equal distance to point C. He then walked in a direction CD, perpendicular to AC until the spike and the ship could be seen in a straight line. Can you fill in the details and show how congruent triangles play a role in determining the value of AE? A solution to this problem will be presented in Chap. 7. **Note:** You will have to use the equality, $\angle DBC = \angle EBA$ which will be proved in Sec. 8.6.

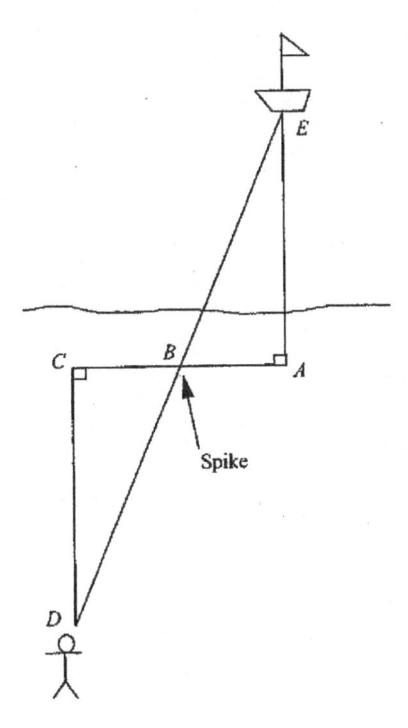

Fig. 5.4. Eudoxes' method for finding the distance to a ship from land.

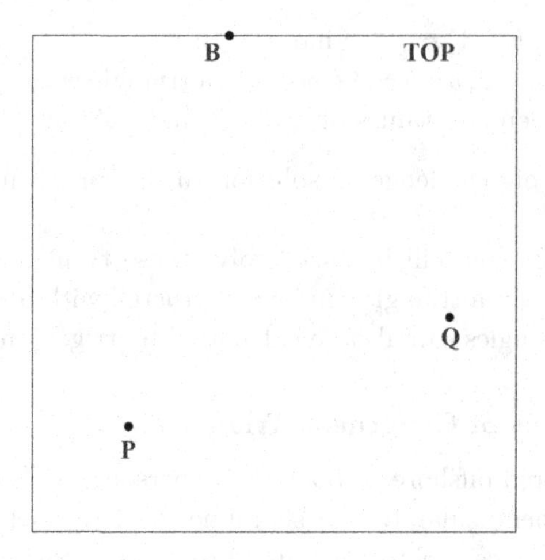

Fig. 5.5. Shortest distance between two points in a rectangle via the top edge of the rectangle.

2. Use your ruler to find the shortest distance from point P to point Q, in millimeters, that intersects the top edge of the rectangle in Fig. 5.5. Through which point B on the top edge of the rectangle does the line pass? Congruent triangles play an important role in the solution to this problem. Try to figure how congruent triangles can help you locate B, or simply use trial and error. A solution to this problem will be presented in Problem 9 of Chap. 7. It is helpful to enlarge Fig. 5.5 or recreate it by placing by recreating points P and Q on a piece of graph paper.

CHAPTER 6

INTRODUCTION TO TRIGONOMETRY

6.1. Introduction

Chapter 5 was devoted to the study of congruent triangles and the determination of angles and sides of a triangle using compass, straightedge, ruler and protractor. Trigonometry is a subject that was developed to determine the edge lengths and angles of a triangle without the use of construction. However, we will find that whenever a triangle can be constructed from partial information about it, the triangle can also be determined by trigonometry. What is so special about a triangle? We will see in Chap. 14 that all polygons can be subdivided into triangles and in turn, closed curves of a more general nature can be approximated by polygons. Furthermore, the triangle distinguishes itself among all polygons as a rigid structure. Any other polygon can be deformed and even bent into three dimensions; only the triangle maintains its rigidity as we will see in Chap. 11. Furthermore, we will see in Chap. 18 that the transformations of Euclidean geometry that result in rigid body motions are completely determined by their action on the three vertices of a triangle. This chapter is devoted to the study of trigonometry applied to triangles.

6.2. The Trigonometry of a Right Triangle

Trigonometry is a subject that begins with a right triangle ΔABC shown in Fig. 6.1.

Given an internal angle θ of the triangle and the length of an edge, the triangle can be constructed with compass, straightedge, ruler and protractor.

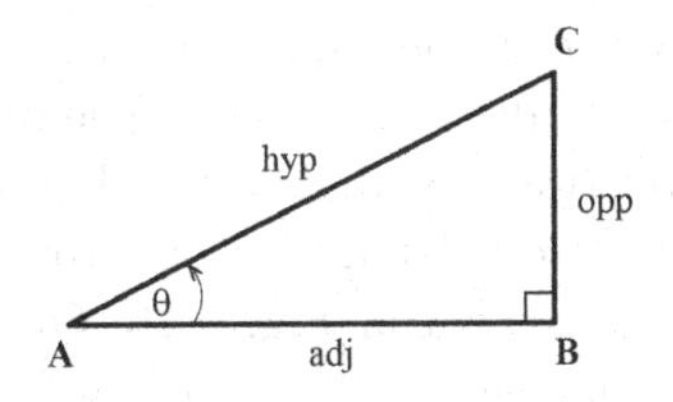

Fig. 6.1. A right triangle.

Problem 1: Let $\angle A = 90\,\text{deg.}$, $\angle B = 30\,\text{deg.}$, and $c = AB = 2$ units, construct ΔABC.

The sides adjacent and opposite to θ and the hypotenuse are abbreviated: adj, opp and hyp, respectively.

Define the ratios: sine (sin), cosine (cos) and tangent (tan) by,

$$\sin\theta = \frac{\text{opp}}{\text{hyp}}, \quad \cos\theta = \frac{\text{adj}}{\text{hyp}}, \quad \tan\theta = \frac{\text{opp}}{\text{adj}}. \tag{6.1a}$$

Remark:

$$\tan\theta = \frac{\text{opp}}{\text{adj}} = \frac{\frac{\text{opp}}{\text{hyp}}}{\frac{\text{adj}}{\text{hyp}}} = \frac{\sin\theta}{\cos\theta}. \tag{6.1b}$$

The inverses of these ratios are called secant (sec), cosecant (csc) and cotangent (cot):

$$\sec\theta = \frac{\text{hyp}}{\text{adj}}, \quad \csc\theta = \frac{\text{hyp}}{\text{opp}}, \quad \cot\theta = \frac{\text{adj}}{\text{opp}} = \frac{\cos\theta}{\sin\theta}. \tag{6.1c}$$

In principle, if the triangle is accurately drawn and the measurements are made arbitrarily precise, these ratios can be exactly determined. In practice, they can be determined only to about two significant figures by measurement. However, using techniques from calculus, they can be determined to an unlimited number of decimal places. Your calculator has been programmed to use the calculus approach, and it generally reproduces trigonometric functions to eight or ten decimal places.

Some of you may have encountered in high school a mneumonic devise to help memorize these ratios:

SOHCAHTOA

which translates to: "Sine is Opposite over Hypotenuse"; "Cosine is Adjacent over Hypotenuse"; "Tangent is Opposite over Adjacent."

What if the triangle is scaled up by doubling all sides? Do you see that the ratios are unchanged? The same would hold if the triangle was scaled up or down by any scale factor. Therefore, sine, cosine, tangent do not depend on the size of the triangle but only on the internal angle θ. In other words, we have defined six trigonometric functions that depend only on θ where $0 < \theta < 90$ deg.

The values of these functions can be found using a calculator. However, there are three values of θ for which the trigonometric functions can be found without the use of a calculator: $\theta = 30\,\text{deg.}$, $60\,\text{deg.}$ and $45\,\text{deg.}$ The angles of $30\,\text{deg.}$ and $60\,\text{deg.}$ are derived from an equilateral triangle as shown in Fig. 6.2a for an equilateral triangle with edge 2 units. A perpendicular from C is dropped to the base AB results in the so-called 30,60,90-triangle shown in Fig. 6.2b. On the other hand, a 45-deg. angle is the angle between the diagonal and side of a half-square shown in Figs. 6.2c and 6.2d.

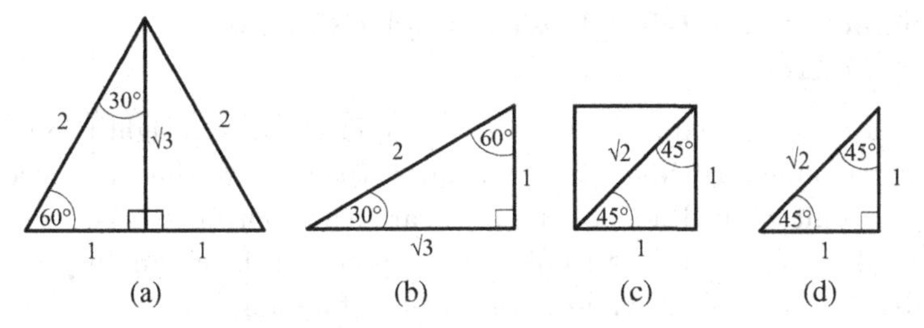

Fig. 6.2. (a) Equilateral triangle cut into two 30,60,90-triangles; (b) A 30,60,90-triangle; (c) A square cut into two 45,45,90-triangles; (d) A 45,45,90-triangle.

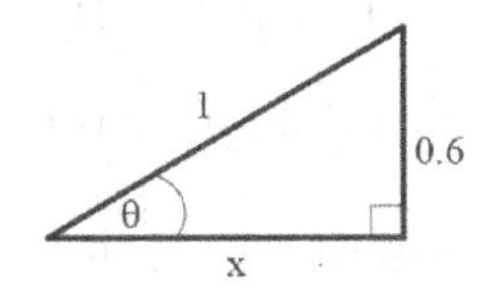

Fig. 6.3. Given that $\sin\theta = 0.6$, find the other trig functions of the angle.

Example: From Figs. 6.2b and 6.2d, and Eq. (6.1a), we see that:

$$\cos 60 = \frac{1}{2}, \quad \sin 30 = \frac{1}{2}, \quad \cos 30 = \frac{\sqrt{3}}{2}, \quad \tan 60 = \sqrt{3}, \quad \sin 45 = \frac{1}{\sqrt{2}}, \quad \tan 45 = 1.$$

Remark: The angles 30 deg., 60 deg. and 45 deg. were the basis of ancient sacred geometry and architecture, and it is noteworthy that these angles formed the basis of the triangle-circle and square-circle grids of Chap. 1.

6.3. Given the Value of One Trigonometric Function to Derive the Others

Given the value of one of the trigonometric functions, you are able to determine the others. For example, if $\sin\theta = 0.6$, the corresponding right triangle is shown in Fig. 6.3. The third side x can be determined by the Pythagorean Theorem, i.e.,

$$x = \sqrt{1^2 - (0.6)^2} = 0.8.$$

Therefore, using the definitions in Eq. (6.1),

$$\sin\theta = 0.6, \quad \cos\theta = 0.8, \quad \tan\theta = \frac{0.6}{0.8}, \quad \csc\theta = \frac{1}{0.6}, \quad \cot\theta = \frac{0.8}{0.6}, \quad \sec\theta = \frac{1}{0.8}.$$

Problem 2: Determine the values of the other trigonometric functions when,

(a) $\tan\theta = 2$,
(b) $\sin\theta = 1/2$,
(c) $\cos\theta = 2/3$.

6.4. Trigonometry of a Triangle with Angles Greater than 90 Degrees

What if θ is greater than or equal to 90 deg.? We no longer have a right triangle. Can we still define the trigonometric functions? The answer is yes. We simply embed the right triangle within a unit circle drawn in an (x, y)-cartesian coordinate system as shown in Fig. 6.4. The radius of the circle is 1 unit while a point on the circumference at angle θ has coordinates $(\cos\theta, \sin\theta)$. Therefore, the cosine of an angle is the x-coordinate of a point on the unit circle at angle θ while the sine of the angle is the y-coordinate. Also by the Pythagorean Theorem, it follows that,

$$\cos^2\theta + \sin^2\theta = 1. \tag{6.2}$$

To compute the tangent, simply divide the y-coordinate by the x-coordinate of the point at angle θ.

If θ is equal to or greater than 90 deg., i.e., $\theta \geq 90$, we no longer have a right triangle, but the sine and cosine functions are still defined as the (x, y)-coordinates of a point on the unit circle at angle θ. Each angle is defined by a ray or half-line that extends from the origin. For example, the positive x-axis represents the 0-deg. ray, while the positive y-axis is the 90-deg. ray, and the negative x-axis is the 180-deg. ray, etc. Angles measured from the 0-deg. ray in a counterclockwise direction are considered to have positive values while angles measured clockwise from the 0-deg. ray have negative values, e.g., the negative y-axis can be considered to have the value $+270$ deg. or -90 deg. Notice that the 0-deg. ray can also be considered as the 360-deg. ray, and likewise, any ray has an alternate value by adding multiples of 360 deg. to it. (Why?)

Using this definition, we can now find the following values of sine and cosine:

$$\sin 90 = 1, \quad \sin 180 = 0, \quad \cos 90 = 0, \quad \cos 180 = -1, \quad \cos 270 = 0, \quad \tan 180 = 0.$$

You can see that the sign $(\pm, \pm) \equiv \mathrm{sgn}(\cos, \sin)$ of sine and cosine depends upon in which quadrant the θ-ray is located as listed in Table 6.1 and Fig. 6.4.

The value of angle θ can now be related to an angle within a right triangle by the following:

If θ is in the following quadrants:

Quadrant 2 $(-, +)$: Drop a perpendicular from the point on the unit circle at the θ-ray to the negative x-axis as shown in Fig. 6.5a. The right triangle is the triangle

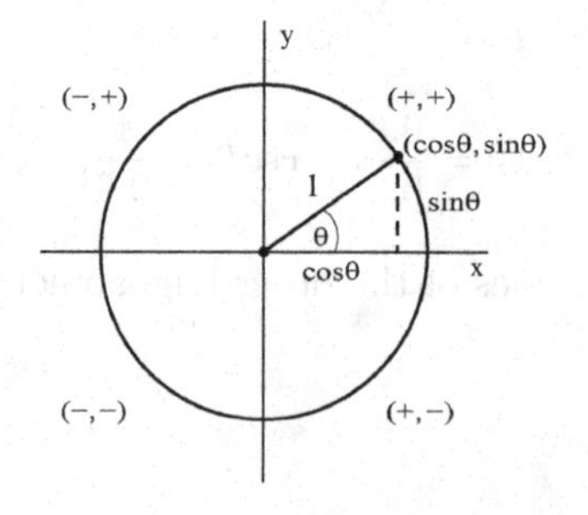

Fig. 6.4. Extending the domain of a trig function beyond right triangles.

Table 6.1.

Quadrant 1: $(0 \le \theta < 90)$: sgn(cos, sin) $\equiv (+,+)$
Quadrant 2: $(90 \le \theta < 180)$: sgn(cos, sin) $\equiv (-,+)$
Quadrant 3: $(180 \le \theta < 270)$: sgn(cos, sin) $\equiv (-,-)$
Quadrant 4: $(270 \le \theta < 360)$: sgn(cos, sin) $\equiv (+,-)$

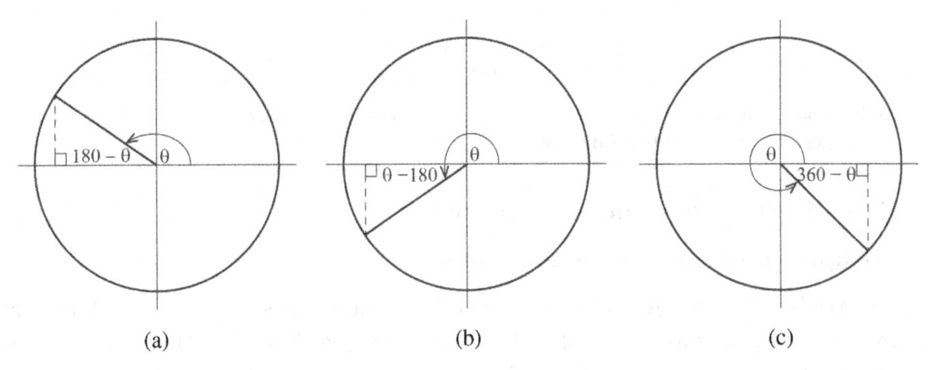

Fig. 6.5. Computing the trig functions of angles in (a) quadrant 2; (b) quadrant 3; (c) quadrant 4.

with the acute angle $(180 - \theta)$ deg. Determine the cosine and sine of this angle, and then append the sign to its value according to Table 6.1.

Quadrant 3 $(-,-)$: Drop a perpendicular from the point on the unit circle at the θ-ray as shown in Fig. 6.5b. The right triangle is the triangle with acute angle $(\theta - 180)$ deg. Determine the sine and cosine of this angle, and then append the sign to its value according Table 6.1.

Quadrant 4 $(+,-)$: Drop a perpendicular from the point on the unit circle at the θ-ray to the positive x-axis as shown in Fig. 6.5c. The right triangle now has acute angle $(360 - \theta)$ deg. Determine the sine and cosine of this angle, and then append the sign to it according to Table 6.1.

Examples:

$$\cos 150 = -\cos 30 = -\frac{\sqrt{3}}{2}$$

$$\sin 225 = -\sin 45 = -\frac{1}{\sqrt{2}}$$

$$\tan 300 = \frac{\sin 300}{\cos 300} = \frac{-\sin 60}{+\cos 60} = \frac{-\sqrt{3}/2}{1/2} = -\sqrt{3}.$$

Problem 3: Determine

(a) $\sin 150$,
(b) $\tan 225$,
(c) $\tan 120$,
(d) $\cos 330$.

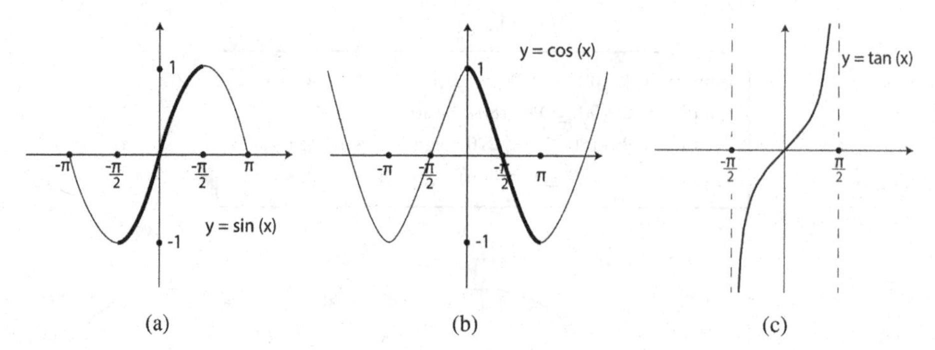

Fig. 6.6. Trig functions as periodic functions. (a) sine; (b) cosine; (c) tangent. The bold faced curves define the graphs of the inverse trig functions.

6.5. Periodic Functions and their Inverses

6.5.1 Graphs of trigonometric functions

Trigonometry began with the measurement of angles and sides of a triangle. Now that we have extended it beyond the realm of triangles, we see that all of the trigonometric functions are periodic with period 360 deg. As angle θ rotates from a given point on the unit circle by 360 deg. we return to the same point. Graphs showing one period of sin, cos, and tan are shown in Fig. 6.6. Notice that the tangent has period 360 deg., but it also has period 180 deg. As a result of their periodic motion, trigonometric functions play an important role in studying the motion of vibrating strings in music, rotary motion of mechanical equipment, the beating of the heart and countless other mathematical models of periodic motion.

6.5.2 Inverse trigonometric functions

If we are given the value of a trigonometric function, how do we determine its angle? For example if $\sin\theta = 0.6$ what is the value of θ? To solve this problem, go to your calculator and input the value of the sine to the $\sin^{-1}$ or arcsin function and you find that $\theta = 36.87$. However, because of the periodicity of the sine function, you will notice from Fig. 6.6 that there are many values of θ that satisfy $\sin\theta = 0.6$, but they all differ by some multiple of 360 deg. What your calculator has done is to compute the principal value of the arcsin function. The principal value of arcsin is derived from the darkened portion of Fig. 6.6a which contains only a portion of the sine curve. On this portion of the curve, you will notice that there is a unique value of θ for each given value of sine. The darkened portions of Figs. 6.6b and 6.6c contain the curves defining the principal values of the cosine and tangent functions.

For problems involving a triangle, we are interested in values of θ in the upper half of the unit circle, i.e., $0 < \theta < 180$ deg. Sine and cosine have values in the interval $[-1, 1]$. Given a number in this interval, e.g., 0.5, there is an important difference between computing $\cos^{-1} 0.5$ and $\sin^{-1} 0.5$. If I input $\cos^{-1} 0.5$ to my calculator, the calculator responds with $\theta = 60$ deg. This result is illustrated in Fig. 6.7a where 0.5 is placed on the y-axis since cosine is the x-coordinate of a point on the unit circle at

60 deg. while $\cos^{-1}(-0.5)$ gives a value of 120 deg. In this way, each value of y in the interval $[-1, 1]$ gives rise to a unique value of θ in the interval $0 \leq \theta \leq 180$.

Now let us consider $\sin^{-1} 0.5$. If I input this value, my calculator responds with $\theta = 30$ deg. Figure 6.7b illustrates this value where $\theta = 30$ deg. corresponding to 0.5 placed on the y-axis since sine is the y-coordinate of a point on the unit circle corresponding to angle 30 deg. However, notice that $\theta = 150$ deg. also has a y-value of 0.5. In other words, there is a second angle in the interval $0 < \theta < 180$ corresponding to a sine value of 0.5. You can also see that 150 deg. = 180 – 30 deg. In general, if you wish to find the angle θ corresponding to $\sin\theta = y$ for $-1 < y < 1$ there are two values of θ, namely, $\theta_1 = \sin^{-1} y$ and $\theta_2 = 180 - \sin^{-1} y$. One angle will be acute while the other is obtuse. In any problem you must determine which of these values makes sense.

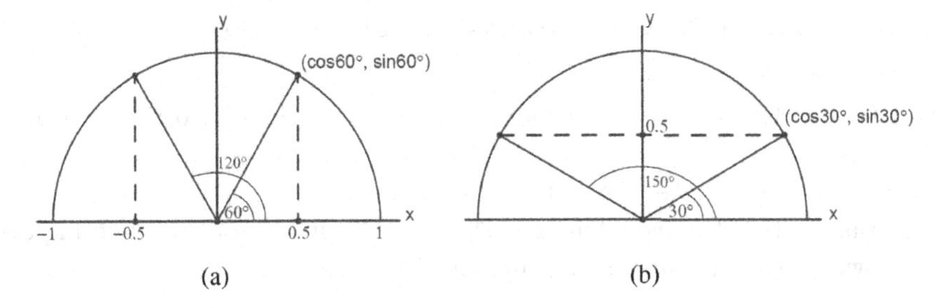

Fig. 6.7. (a) Inverse cosine for obtuse angles; (b) inverse sine for obtuse angles.

Problem 4: Find all values of $0 \leq \theta \leq 180$ satisfying:

(a) $\cos\theta = 0.4$,
(b) $\cos\theta = -0.7$,
(c) $\sin\theta = 0.4$.

6.6. Application of Trigonometry to Compute the Angles and Sides of a Triangle Given Partial Information About the Triangle

The subject of *trigonometry* has been developed to determine side lengths and angles of triangles without the need to construct them. Whenever partial information about a triangle enables you to construct the triangle, then all sides and angles can be determined by trigonometry. We find, in general, that three pieces of information about the triangle must be specified. These can be basic elements or other features of the triangle such as altitudes or medians. For right triangles we can generally solve for the basic elements using SOHCAHTOA. If the triangle is not a right triangle then we may need to use two additional laws: the *Law of Cosines* and the *Law of Sines.* We will first illustrate this for a right triangle.

6.6.1 Right triangles

If the triangle is a right triangle, then given an angle and a side, the remaining sides and angles can be determined.

Example: Given hypotenuse $b = AC = 3$ units in Fig. 6.8a and $\angle A = 30$ deg., solve for $BC = a$.

Let $a = x$, $\sin 30 = \frac{x}{3}$, $\frac{1}{2} = \frac{x}{3}$. Therefore, $x = \frac{3}{2}$.

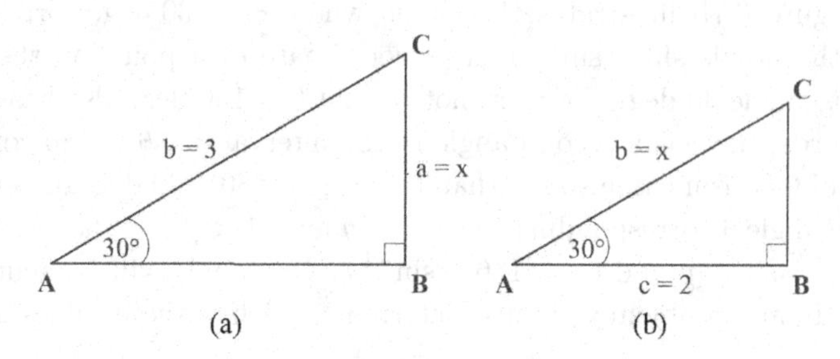

Fig. 6.8. (a) (b) two right triangles to solve for an unknown side x.

Example: Given $AB = c = 2$ units and $\angle A = 30\,\text{deg}$. in Fig. 6.8b, solve for the hypotenuse, $AC = b$.

Let $b = x$, $\cos 30 = \frac{2}{x}$, $\frac{\sqrt{3}}{2} = \frac{2}{x}$. Therefore, $x = \frac{4}{\sqrt{3}}$.

If the triangle is not a right triangle, then usually it helps to use two important relations known as the *Law of Cosines* and the *Law of Sines.*

Problems: Given the following information about a right triangle, construct the triangle with compass, straightedge and protractor. Choose *four* graph paper boxes as the unit, and determine the unknown sides and angles by measurement. Then check your results by determining the values of these sides and angles using trigonometry.

5. Let $\angle A = 50$ deg., $\angle B = 90$ deg., $b = 2$. Determine c.
6. Let $\angle A = 40$ deg., $\angle B = 90$ deg., $a = 1$. Determine b
7. Let $\angle A = 30$ deg., $\angle B = 90$ deg., $b = 2$. Determine a.
8. Let $\angle A = 30$ deg., $c = 3$, $h_c = 1$. Determine a, b and $\angle B$.
9. Let $a = 2$, $c = 3$, $h_c = 1$. Determine b and $\angle A$. Note: There are two distinct triangles that can be constructed and their angles and sides can be found from the trigonometry of a right triangle.

6.6.2 The law of cosines

For triangle ΔABC (see Fig. 6.9),

$$c^2 = a^2 + b^2 - 2ab\cos C. \tag{6.3}$$

In other words, if you know two sides, a, b and the included angle C, you can solve for the side opposite the angle. For a proof, see the Appendix 15A.

6.6.3 Law of sines

For ΔABC (see Fig. 6.9),

$$\frac{a}{\sin A} = \frac{b}{\sin B} = \frac{c}{\sin C}. \tag{6.4}$$

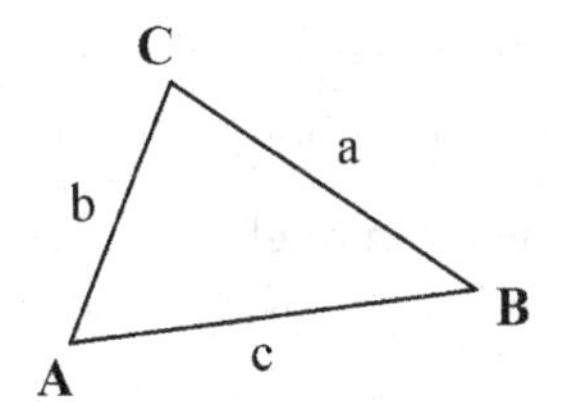

Fig. 6.9. A reference triangle.

Here the sides of a triangle are related to their opposite angles. For a proof, see Appendix 6A.

6.6.4 SAS (see Fig. 6.10)

Given $b = 1$, $c = 2$ and $\angle A = 120\,\text{deg.}$, i.e., SAS, find a.

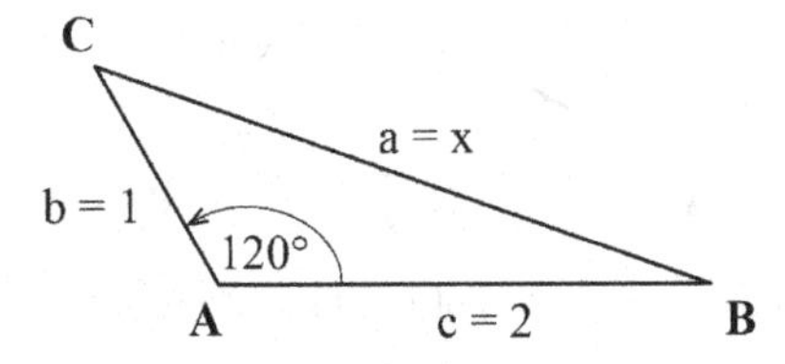

Fig. 6.10. Application of the Law of Cosines to SAS.

Note that since $\angle A = 120\,\text{deg.}$, which is in Quadrant 2, you can use Fig. 6.5a to relate this angle to the acute angle 60 deg. and the 30,60,90-triangle in Fig. 6.2b to see that

$$\cos 120 = -\cos 60 = -\frac{1}{2} \tag{6.5a}$$

and

$$\sin 120 = \sin 60 = \frac{\sqrt{3}}{2}. \tag{6.5b}$$

Let $a = x$. From the Law of Cosines and using Eq. (6.5a),

$$x^2 = 1 + 4 - 2(1)(2)\cos 120 = 5 - 4\left(-\frac{1}{2}\right) = 7. \tag{6.6a}$$

Therefore, $x = a = \sqrt{7}$.

Now that you know the length of the three sides of the triangle and one angle, you can find $\angle B$ by using the Law of Sines.

$$\frac{a}{\sin A} = \frac{b}{\sin B} \quad \text{or} \quad \frac{\sqrt{7}}{\sin 120} = \frac{1}{\sin B}.$$

Cross multiplying, and using Eq. (6.56b),

$$\sin B = \frac{\sin 120}{\sqrt{7}} = \frac{\sqrt{3}/2}{\sqrt{7}} = \frac{\sqrt{3}}{2\sqrt{7}} = 0.3273. \tag{6.6b}$$

Therefore, $\angle B = \sin^{-1} 0.3273 = 19.11$ deg. or $\angle B = 160.89$ (making use of Sec. 6.5.2). By constructing the triangle it is obvious that value of $\angle B = 19.11$ deg.

Remark: Given the sine or cosine of an angle, that angle can be found by using the inverse trigonometric functions $\sin^{-1}$ and $\cos^{-1}$ on your calculator.

Since the angles A, B, C sum up to 180 deg., $\angle C = 180 - (120 + 19.105) = 40.895$ deg.

Check this result by measuring side a and $\angle B$ with a ruler and protractor.

6.6.5 ASA (see Fig. 6.11)

Given $\angle A = 30$ deg, $c = 2$ units, $\angle B = 45$ deg. i.e., ASA, find a and b.

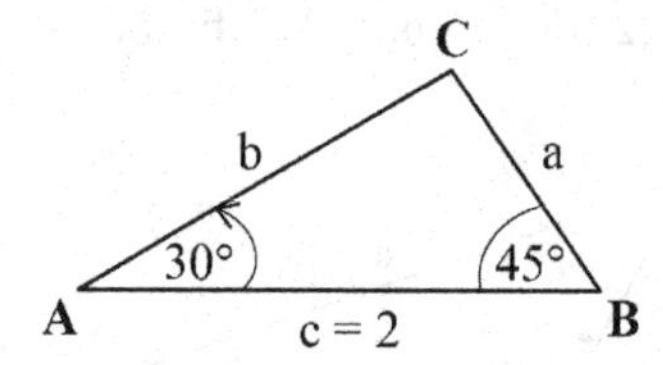

Fig. 6.11. Application of the Law of Sines to ASA.

Let $a = x$:

$$\angle C = 180 - (\angle A + \angle B) = 180 - 75 = 105 \text{ deg.}$$

From the Law of Sines:

$$\frac{a}{\sin A} = \frac{c}{\sin C} \quad \text{or} \quad \frac{x}{\sin 30} = \frac{2}{\sin 105}$$

$$x = \frac{2 \sin 30}{\sin 105} = \frac{2(1/2)}{0.9659} = 1.035$$

$$a = 1.035.$$

Repeating this for b yields

$$b = 1.464.$$

Problems: Given the following partial information about a triangle, construct the triangle with compass, straightedge and protractor. Choose an appropriate number of graph paper boxes as the unit, and determine the unknown sides and angles by measurement. Then check your results by determining the values of these sides and angles using the *Law of Cosines* and *Law of Sines*.

10. Let $\angle A = 150$ deg., $b = 1$, $c = 2$. Determine c, $\angle B$.
11. Let $a = 6$, $b = 9$, $c = 13$. Determine $\angle A$, $\angle B$, $\angle C$.
12. Let $\angle A = 60$, $c = 2$, $\angle B = 45$. Determine a, b.

6.7. Calculation of SSA

We saw in Sec. 5.3.3 that more than one triangle can be constructed given SSA. In this section, we will analyze the various possibilities of SSA.

Consider $c = 2$, $a = 1$ and three values of $\angle A$: 20 deg., 30 deg. and 40 deg. (i.e., SSA). Since we know that $a = 1$, the possible positions for side BC lie on the semicircle shown in Fig. 6.12a for each of the three given angles. Let us see what the Law of Cosines has to say about the unknown side $b = x$ for each of the three given values of $\angle A$.

a. $\angle A = 20$ deg. (see Fig. 6.12b). From the Law of Cosines:

$$1^2 = x^2 + 2^2 - 2(2)(x)\cos 20. \tag{6.7}$$

Rewriting Eq. (6.7),

$$x^2 - 4x\cos 20 + 3 = 0. \tag{6.8}$$

Since $\cos 20 = 0.94$, Eq. (6.8) becomes,

$$x^2 - 3.76x + 3 = 0. \tag{6.9}$$

Notice that this is a quadratic equation to solve for $a = x$. Remember that to solve a general quadratic equation,

$$Ax^2 + Bx + C = 0 \tag{6.10}$$

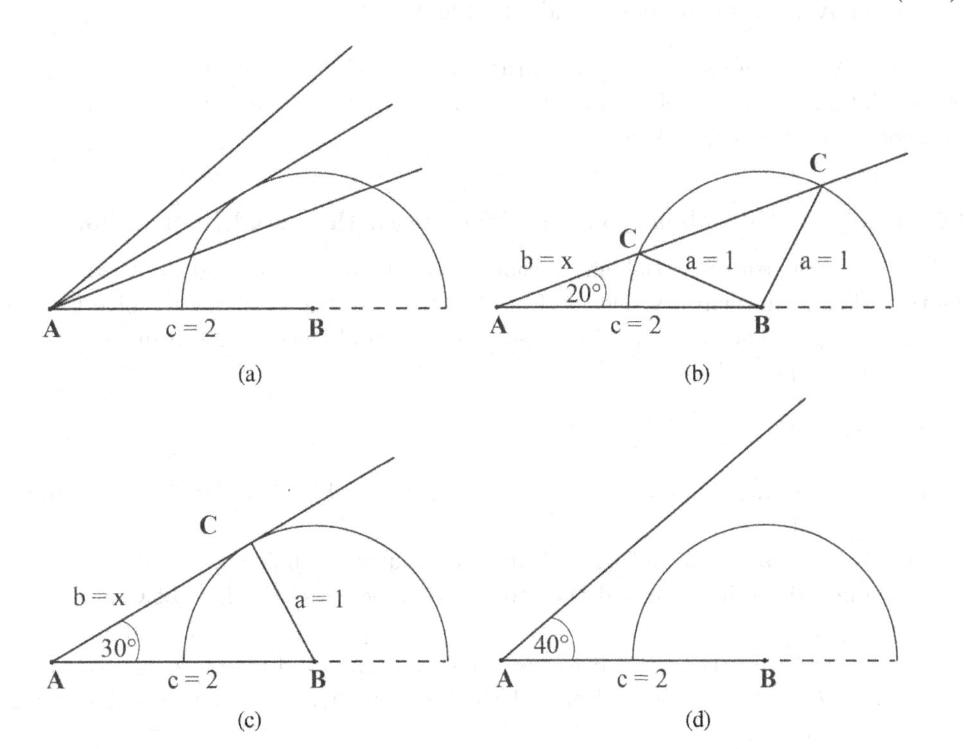

Fig. 6.12. (a) Three cases of SSA; (b) Case 1 — two solutions; (c) Case 2 — one solution; (d) Case 3 — no solutions.

you can use the quadratic formula:

$$x = \frac{-B \pm \sqrt{B^2 - 4AC}}{2A} \tag{6.11}$$

to solve for $b = x$ in Eq. (6.9) which we identify from Eq. (6.7) that $A = 1$, $B = -3.76$, and $C = 3$. Therefore $x = 2.61$ and $x = 1.15$, resulting in the two solutions shown in Fig. 6.12b.

b. Let $\angle A = 30$ deg. (see Fig. 6.12c) where $\cos 30 = \frac{\sqrt{3}}{2}$.
This results in the quadratic equation,

$$x^2 - 2\sqrt{3}x + 3 = 0. \tag{6.12}$$

To solve Eq. (6.12), we identify from Eq. (6.10) that $A = 1$, $B = -2\sqrt{3}$, and $C = 3$ resulting in $b = x = \sqrt{3}$. For this case we obtained a single solution because the expression under the square root sign in Eq. (6.9) equals zero, (check this) validating Fig. 6.12c.

c. Let $\angle A = 40$ deg. (see Fig. 6.12d)

We again obtain a quadratic equation, but this time the equation has no solutions because the expression under the square root sign is negative, and negative numbers have no real-valued square roots, validating Fig. 6.12d.

Remark: We have for this example a situation in which SSA leads to either two, one, or no solutions constructible from this partial information. We clearly see why SSA does not guarantee congruence.

6.8. Additional Problems to Solve Triangles with Partial Information

In Sec. 5.4, we presented partial information about a triangle and asked you to construct the triangle and measure its sides and angles. Now use trigonometry to solve for the sides and angles. The constructions to some of the problems are given in the figures accompanying the problems.

Problems:

13. Let $\angle A = 45$ deg., $c = 3$ and $h_c = 1$, as shown in Fig. 5.3, Sec. 5.4. Determine $a, b, \angle B$ and $\angle C$.
14. Supply two sides and an included angle of your own choice, i.e., SAS, and then determine the third side and the other two angles. Use the Law of Cosines and Law of Sines.
15. Supply ASA and then determine the remaining angle and the other two sides.
16. Let $a = 2$, $\angle A = 40$ deg. and $h_c = 1$ as shown in Fig. 6.13. Determine the other two sides and angles.
17. Let $a = 4, b = 3$ and $h_c = 2$ as shown in Fig. 6.14 with its construction lines. Determine side c and all three angles. Notice that four triangles are possible although two of them are congruent to the other two.

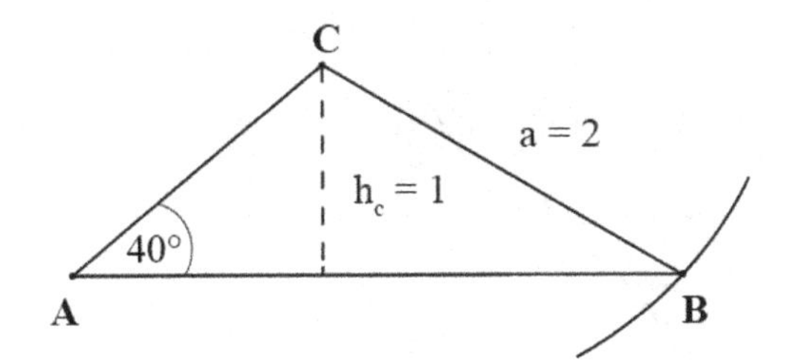

Fig. 6.13. Use trig to solve problem 16.

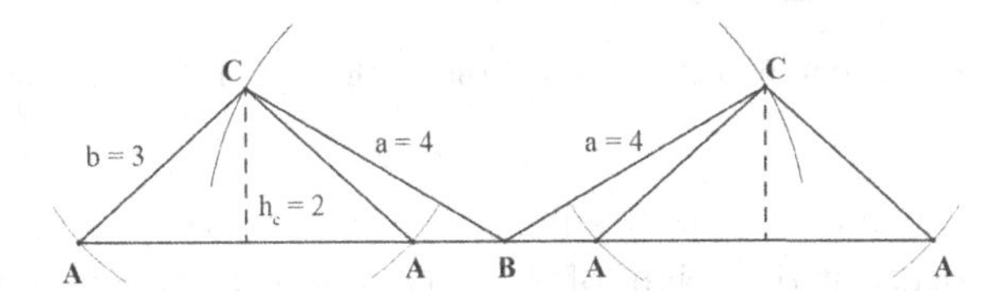

Fig. 6.14. Use trig to solve problem 17.

18. Let $c = 4, h_c = 1$ and $m_c = 2$ as shown in Fig. 6.15 with construction lines. Determine the sides b, c and all three angles. Notice that $\angle C = 90\,\text{deg}$. (Why?).
19. Given: $\angle A = 30\,\text{deg}, \angle B = 60\,\text{deg}$, $a = 2$. Determine b and c.

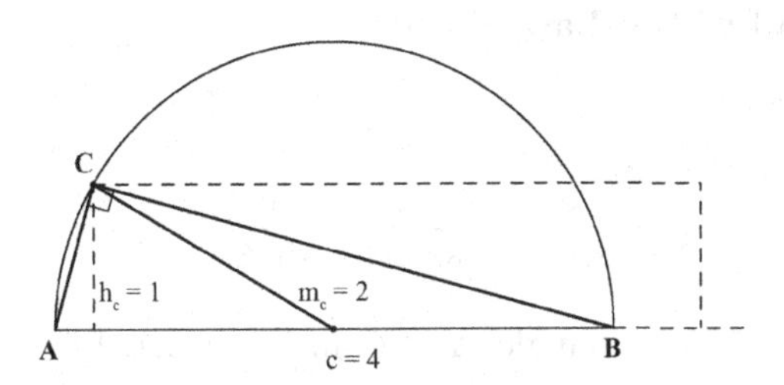

Fig. 6.15. Use trig to solve problem 18.

6.9. An Application of Trigonometry to Finding the Radius of Earth

Simple geometry and trigonometry can be used to find the radius of Earth. Imagine that you are standing atop a mountain on an island in the middle of the ocean. The mountain has height, h. You site the horizon and find that your line of sight must decrease from a direction straight ahead by an angle of declination ϕ as shown in Fig. 6.16. Since the line of sight to the horizon is tangent to Earth, a right triangle is set up between your position at the top of the mountain, the position of the horizon and the center of Earth.

Problem 20: If the mountain has height, $h = 986$ meters, and the declination angle, $\phi = 1$ deg., using Fig. 6.16, the Pythagorean Theorem and trigonometry, compute the radius of Earth in miles, and compare it with the actual value of 3963 miles.

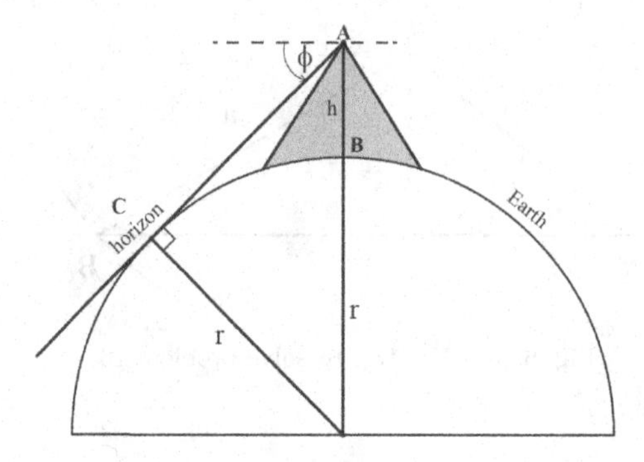

Fig. 6.16. Using trig to find the radius of the earth or the height of a mountain.

Problem 21: Now assume that the radius of Earth is 3963 miles, and the angle of declination is still 1 deg. but the height of the mountain is unknown. Solve this triangle for h, the height of the mountain, and compare it with the above value of 986 meters.

Remark: Visualize yourself traveling in an airplane and continuously sighting the horizon. By measuring the angle of declination in real time, your altitude at any moment can be known.

Appendix 6A. Proof of the Law of Sines

Theorem: (Law of Sines)

$$\frac{a}{\sin A} = \frac{b}{\sin B} = \frac{c}{\sin C}.$$

Proof: Consider the reference triangle ΔABC in Fig. 6A.1. Let h_c be the altitude to side c:

$$\sin A = \frac{h_c}{b}, \quad \sin B = \frac{h_c}{a}.$$

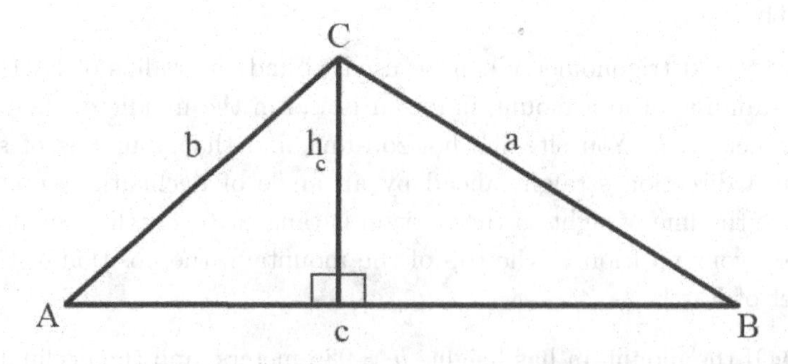

Fig. 6A.1. Proof of the Law of Sines.

Therefore,

$$h_c = b \sin A = a \sin B \quad \text{or} \quad \frac{a}{\sin A} = \frac{b}{\sin B}.$$

The same argument can extend to $\angle C$ and side c in which case,

$$\frac{a}{\sin A} = \frac{b}{\sin B} = \frac{c}{\sin C}.$$

QED

CHAPTER 7

THE ART OF PROOF

7.1. Introduction

Geometry begins with a set of *axioms* which are apriori statements considered to be true which lead to other true statements called *theorems*. A proof can be thought of as a step by step process leading from hypothesis, assumed to be true, to conclusion. This path from hypothesis to conclusion consists of axioms; definitions; other theorems previously proven; and laws of logic, i.e.,

$$\text{Hypotheses} + \text{Axioms} + \text{Definitions} + \text{Other theorems} + \text{Laws of logic}$$
$$\Rightarrow \text{Conclusion}$$

Thus, all proofs take the form of a sequence of statements each of which is true, leading from hypothesis to conclusion. When this process is carried out, the result will be a new theorem which will be another true statement within the system governed by the original axioms. This process lies at the foundation of mathematics and science and is called the *deductive method.*

Discoveries in mathematics and science rarely come about as a result of the deductive method. Rather, discoveries are generally the result of observation, experimentation, trial and error, or even dreams. However, any theorem once discovered can, in principle, be validated by subjecting it to the deductive method, and this often leads to a deeper understanding of the theorem.

All geometric propositions or theorems can be expressed in the form:

$$\text{If A then B} \quad \text{or} \quad A \Rightarrow B.$$

Another way to state this is: If A is true then B is true where A and B are statements. The A statement is the *hypothesis* while the B statement is the *conclusion.*

7.2. Three Important Theorems

For example, an important theorem of geometry states:

Theorem 7.1: *If a triangle is isosceles, then the base angles are equal.*

In this proposition, A (the hypothesis) is the statement: "A triangle is isosceles." while B (the conclusion) is the statement: "The base angles are equal."

Theorem 7.2: *If Point C is equidistant from A and B, then C lies on the perpendicular bisector of AB.*

In this proposition, the hypothesis is the statement: "Point C is equidistant from Points A and B." while the conclusion is the statement: "C lies on the perpendicular bisector of AB."

Theorem 7.3. *The perpendicular bisectors of the sides of a triangle meet at a common point which is the center of the circumscribing circle.*

This proposition can be rephrased as follows:

If three lines are the perpendicular bisectors of the sides of a triangle, then the three lines meet at a point which is the center of the circumscribing circle.

The hypothesis is the statement: "Three lines are perpendicular bisectors of the sides of a triangle," while the conclusion is: "The lines meet at a common point which is the center of the circumscribing circle."

Remark: Theorem 7.3 has been rewritten in the form of an "if, then" statement after which the hypothesis and conclusion are identified.

7.3. Converses

The *converse* of a proposition:

$$\text{If A then B} \quad \text{or} \quad A \Rightarrow B$$

is the reverse statement:

$$\text{If B then A} \quad \text{or} \quad B \Rightarrow A.$$

In this respect, there is an important difference between Theorems 7.1 and 7.2 on the one hand, and Theorem 7.3 on the other.

The converse of Theorem 7.1 states:

Theorem 7.1′: *If the base angles of a triangle are equal, then the triangle is isosceles.*

You will be asked in the problems to prove the validity of this statement so it is another theorem.

The converse of Theorem 7.2 states:

Theorem 7.2′: *Any point on the perpendicular bisector of line segment AB is equidistant from A and B.*

You will be asked in the problems to prove this theorem.

The converse of Theorem 7.3 states: "If three lines within a triangle meet at a common point, then the lines are the perpendicular bisectors of the sides of the triangle."

This is certainly *not* true, as we have seen in Sec. 4.3, since the three angle bisectors of a triangle or the three medians of a triangle also meet at a common point, yet they are not perpendicular bisectors of the sides of the triangle. Therefore the converse is false.

Now consider the following true statement: "If a four-sided geometric figure is a square then it has all equal sides."

Its converse is: "If a four-sided figure has all equal sides, then it is a square."

Notice that the converse need not be true. (Why?)

7.4. Equivalence Tautologies and Definitions

A proposition is called an *equivalence tautology* when both the proposition and its converse are true, i.e.,

$$A \Rightarrow B \quad \text{and} \quad B \Rightarrow A.$$

Another way to state such a proposition is:

A is true if and only if B is true.

All *definitions* are equivalence tautologies.

The proper definition of a square can therefore be expressed as follows: If a four-sided figure has four equal sides and four right angles, then the figure is a square.

Notice that this proposition and its converse are true. Therefore, it qualifies as a definition.

We generally define an isosceles triangle as follows:

A triangle is isosceles if and only if it has two equal sides.

In this regard, based on Theorem 7.1 and its converse, Theorem 7.1′, an isosceles triangle can also be defined as follows:

A triangle is isosceles if and only if it has two equal side angles.

7.5. Proof

Any system of mathematics begins with a set of undefined or primitive concepts such as points, lines, planes, and a set of axioms. Euclidean geometry was created by the Greek mathematician, Euclid, who stated five axioms and then went on to prove, from these axioms, enough theorems to fill thirteen books. In describing the system of Euclidean geometry in modern times we no longer rely on Euclid's axioms. Instead, they have been replaced by other axiom systems which are easier to use and avoid certain logical problems [Wal].

The following are three examples of proof. On the left column, we place the sequence of true statements leading from hypothesis to conclusion, while their reasons or justifications are placed in the right column.

I would like to stress the importance of a figure. Although the proof should be independent of the figure, by drawing a figure pertaining to the theorem you wish to

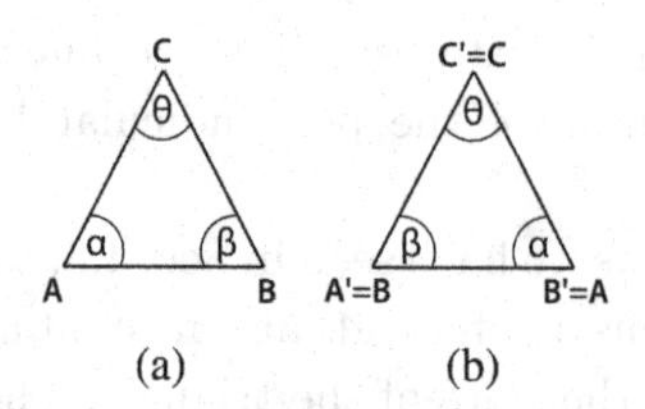

Fig. 7.1. (a) (b) In the proof of Theorem 7.1, isosceles triangle ABC is transformed to A′B′C′ by reflecting it in a mirror through C.

prove, you will be able to develop a strategy for the proof. In other words, the figure helps you to see where you wish to go with the proof. I will now prove Theorems 7.1, 7.2 and 7.3, prefaced by short discussions of the strategies that precede the formal exposition of the proofs.

Theorem 7.1: *If a triangle is isosceles, then the base angles are equal (see Fig.* 7.1).

Strategy for the proof of Theorem 7.1:

To prove Theorem 7.1 that the base angles, α and β, of an isosceles triangle ΔABC are equal, construct a second triangle $\Delta A'B'C'$ which is the reflection of ΔABC, preserving angle θ as shown in Figs. 7.1a and 7.1b in which the base angles α and β change places. Then prove that ΔABC and $\Delta A'B'C'$ are congruent by SAS in which case the base angles are not involved. *Corresponding parts of congruent triangles are equal* then leads to the conclusion that $\alpha = \beta$. We are now in a position to restate this strategy as a formal proof.

Statement	Reason
1. ΔABC is isosceles with AB the base and $\angle C$ the opposite vertex.	Hypothesis.
2. $AC = BC$.	Definition of isosceles.
3. Transform ΔABC to $\Delta A'B'C'$ in such a way that points $\angle CAB = \alpha \to \angle A'B'C' = \alpha$, $\angle ABC = \beta \to \angle C'A'B' = \beta$, $\angle BCA = \theta \to \angle B'C'A' = \theta$.	
4. $AC \to B'C'$, $BC \to A'C'$.	
5. $A'C' = BC = AC$ and $B'C' = AC = BC$.	From Steps 2 and 3.
6. $\angle C = \angle C' = \theta$.	Identity.
7. $\Delta ACB \cong \Delta A'C'B'$.	Congruent by SAS $(5, 6)$.
8. $\alpha = \angle A = \angle A' = \beta$.	Corr. parts of congruent triangles are equal. QED

Remark: QED means *Quod Erat Demonstrandum* or What Was to be Demonstrated.

Remark: In this theorem, $\Delta A'B'C'$ is identical to ΔABC except that vertices A and B have been interchanged. In Figs. 7.1a and 7.1b, we show these as two different triangles.

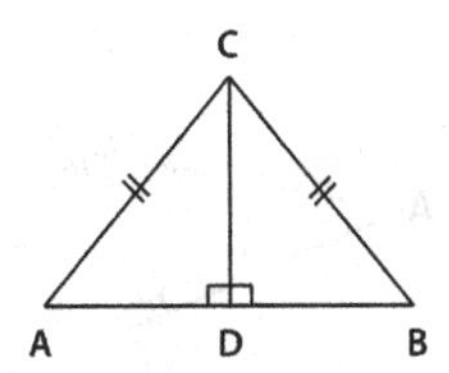

Fig. 7.2. Triangle for the proof of Theorem 7.2.

Theorem 7.2: *If Point C is equidistant from Points A and B, then C lies on the perpendicular bisector of AB (see Fig. 7.2).*

Strategy for the proof of Theorem 7.2:

In Fig. 7.2, a perpendicular bisector CD is dropped from Point C to AB where $CA = CB$ according to the hypothesis. Our objective is to prove that $\Delta ADC \cong \Delta BCD$. Since corresponding parts of congruent triangles are equal, it will then follow that $AD = BD$, i.e., line segment AB has been bisected, validating the conclusion of Theorem 7.2. Now we are in a position to formally prove Theorem 7.2.

Statement	Reason
1. $CA = CB$.	Hypothesis.
2. ΔABC is isosceles with AB the base and $\angle C$ the vertex.	Definition of isosceles.
3. $\angle A = \angle B$.	Base angles of an isosceles triangle are equal.
4. Drop a perpendicular line from C to AB meeting AB at D so that $\angle ADC = \angle CDB = 90$ deg.	Definition of perpendicular.
5. $CD = CD$.	Identity.
6. $\Delta ADC \cong \Delta BCD$.	Congruent by AAS $(3, 4, 5)$.
7. $AD = BD$.	Corresponding edges of congruent triangles are equal.
8. CD is a perpendicular bisector of AB.	Definition of perpendicular bisector.

QED

Remark: Theorem 7.3 justifies Construction 2 of Sec. 4.2.1 to create a perpendicular bisector of a line segment and also for Sec. 4.2.2 to construct a line from a given point perpendicular to a given line.

Theorem 7.3: *The meeting point of the perpendicular bisectors of the sides of a triangle is the center of the circle that circumscribes the triangle (see Fig. 7.3).*

Strategy for proving Theorem 7.3:

A triangle is described by three noncollinear points A, B, C, shown in Fig. 7.3. The perpendicular bisectors of two consecutive sides of this triangle, AB and BC, meet at

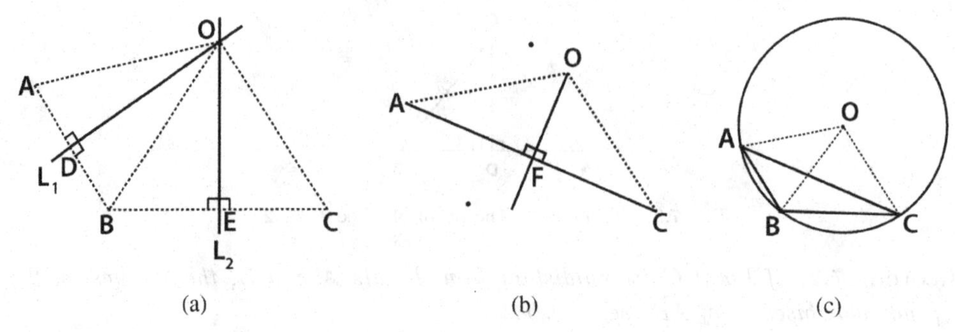

Fig. 7.3. (a) Perpendicular bisectors of two sides of a triangle meet at point O, (b) The perpendicular bisector of the third side is incident to O; (c) O is the center of the circumscribing circle.

Point O which must define the center of the inscribed circle. The strategy will be to show that $AO = BO = CO$ by using congruent triangles. These will then be the radii of the circumscribing circle. Finally, we must show that the perpendicular bisector of the third side AC goes through O; then we will have proved the theorem. Now we are ready to proceed with a formal proof.

Statement	Reason
1. Let A, B, C be three non-collinear points making up the vertices of triangle ΔABC, and l_1 and l_2 be the perpendicular bisectors of AB and BC (see Fig. 7.3a).	Hypothesis.
2. Nonparallel lines l_1 and l_2 intersect at O so that OD and OE are perpendicular bisectors of AB and CD.	Axiom of Euclidean geometry.
3. $AD = DB$.	Definition of bisector.
4. $OD = OD$.	Identity.
5. $\angle ADO = \angle ODB = 90$ deg.	Definition of perpendicular.
6. $\Delta ADO \cong \Delta ODB$.	Congruent by SAS (3, 4, 5).
7. $AO = BO$.	Corresponding parts of congruent triangles are equal.
8. Repeating Steps 3–7, it follows that $BO = CO$.	
9. $AO = BO$.	Follows from Steps 7 and 8.
10. Consider Δ ACO (see Fig. 7.3b), O lies on the perpendicular bisector, OF, of AC so that O is the meeting point of the perpendicular bisectors of ΔABC where $AO = BO = CO$.	Theorem 7.2.
11. O is the center of the circumscribing circle of ΔABC. (see Fig 7.3c)	

QED

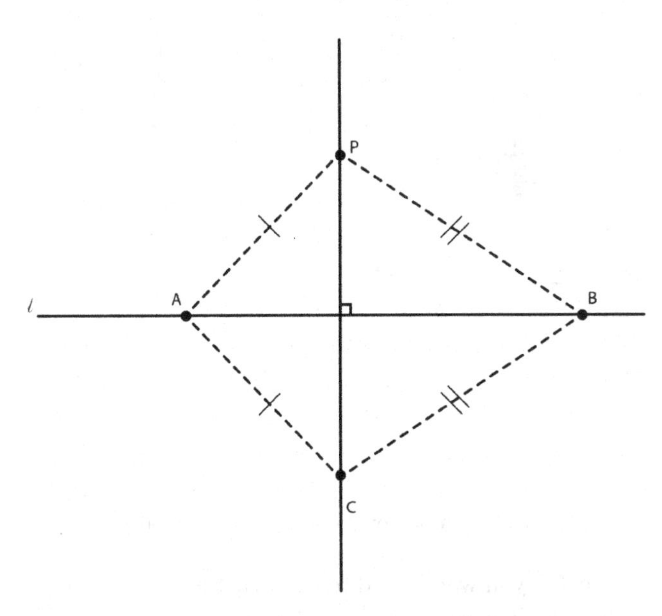

Fig. 7.4. Proof of Method 2 for dropping a perpendicular from point P to line l.

Problems:

Prove the following theorems:

1. **Theorem 7.1′** (Converse of Theorem 7.1).
2. **Theorem 7.2′** (Converse of Theorem 7.2).
3. In Construction 2 of Sec. 4.2.2, Method 2 was an alternative way to construct a line drawn from a point perpendicular to a given line, now redrawn in Fig. 7.4. Justify Method 2 with a proof.
4. In Chap. 5, you were asked to come up with a way to find the distance AE from a point on the shore to a ship moored in the ocean by the method shown in Fig. 5.4 in which the line segment $CD = AE$ where CD is a line segment on dry land. Referring to Fig. 5.4, use the deductive method to prove that $AE = CD$.
   Note: $\angle DBC$ and $\angle EBA$ are referred to as vertical angles. In the proof, you will have to use the fact that vertical angles are equal, i.e., $\angle DBC = \angle EBA$. This will be proven in the next chapter.

5.* A camper wishes to fetch water from a stream and then return to his tent. Find the shortest distance from his starting position P to the stream at B and from the stream to his tent at Q. This is equivalent to the problem in Chap. 5 in which you were asked to find, by trial and error, the shortest distance from point P to the top edge of the diagram at B and then to point Q as shown in Fig. 5.5. Now use the deductive method and Fig. 7.5 to prove that path PBQ is the shortest path. In other words, any other path such as $PB'Q$ would be longer. **Note:** The asterisk means that this is a challenging problem.

Remark: The geometry is similar to a ball thrown against the wall where the angle of incidence equals the angle of reflection. (Prove this!)

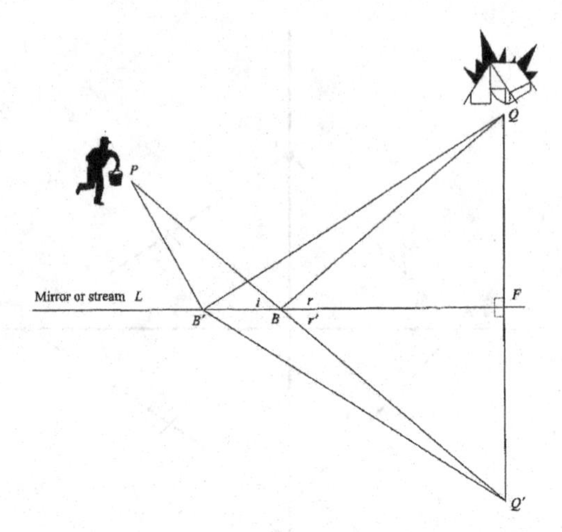

Fig. 7.5. Solution of Problem 11 of Chapter 5.

9. In Problem 9 of Sec. 5.5 you were asked to construct a triangle given: $c = 2$, $\angle A = 60$ deg., and $a + b = 4$. This can be done as shown in Fig. 7.6 by marking off $a + b$ on the 60-deg. ray at E and constructing the perpendicular bisector CD of BE intersecting the 60-deg. ray at C. Then using the converse of Theorem 7.2, we see that CE = BC = a. Let $a = x$ and $b = 4 - x$ and use the Law of Cosines to solve for x.

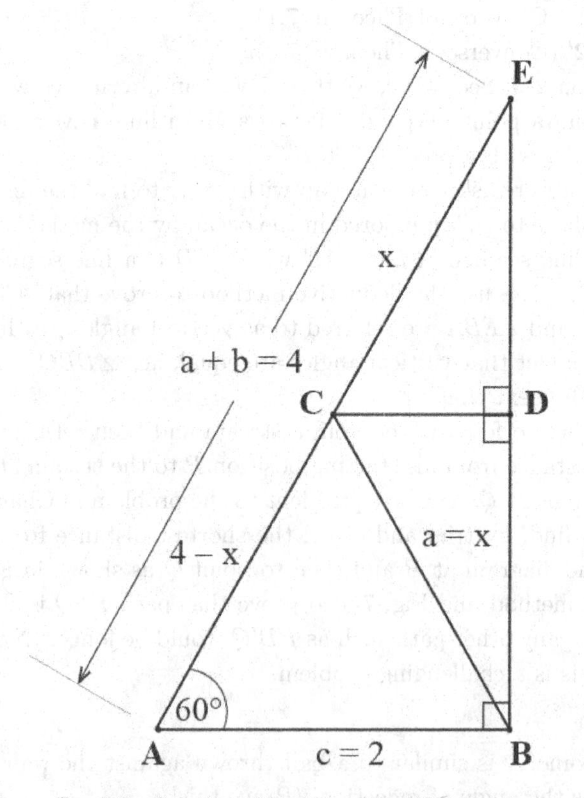

Fig. 7.6. Solution of Problem 9** of Chapter 5.

CHAPTER 8

ANGLE

8.1. Introduction

In Chap. 5, you discovered that the sum of the angles of a triangle is 180 deg. You did this by cutting off the vertices of a triangle and reassembling them to form a straight angle. However, this is not sufficient by the standards of mathematical truth; it requires a proof which will be supplied in Chap. 10. Most of us would be satisfied by the empirical observation of Chap. 5. However, there are other non-Euclidean geometries for which the sum of the angles of a triangle do not add up to 180 deg. For example, consider the geometry on a sphere. One of Euclid's axioms states:

Axiom 8.1: Between any two points in Euclidean geometry a unique line can be drawn.

By a straight line between two points in spherical geometry, we mean the arc of the unique *great circle* containing the points. A great circle is a circle gotten by slicing through the sphere with a plane that contains the center of the sphere. In Fig. 8.1 the sphere is pictured as a globe of the Earth where the longitude lines and

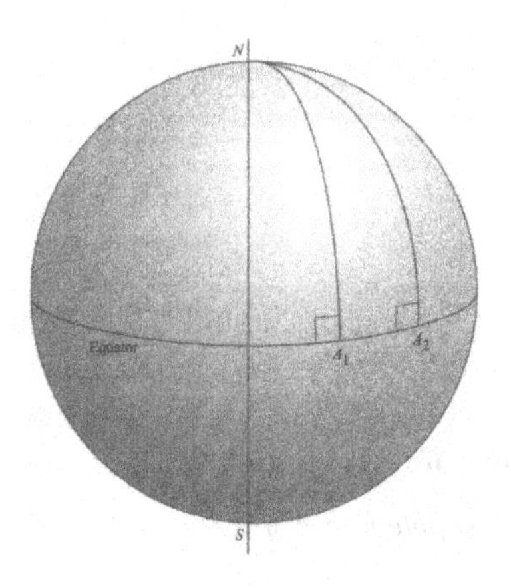

Fig. 8.1. The sum of the angles of a triangle on a sphere is greater than 180 deg.

the equator are great circles. Two longitude lines intersect the equator at right angles forming a spherical triangle A_1, A_2 N, where N refers to the North Pole. Clearly, this triangle has angles that sum to more than 180 deg. For the remainder of this chapter, we will focus on angle in the context of Euclidean geometry.

8.2. Interior Angles of a Triangle

I refer to the sum of the *interior angles* of a triangle, shown in Fig. 8.2, as I_3, where

$$I_3 = \sum_{k=1}^{3} \alpha_k = 180\,\text{deg.} \tag{8.1}$$

Now find the sum of the *interior angles* of a four-sided polygon (quadrilateral). Consider the quadrilateral shown in Fig. 8.3. Subdivide it into two triangles. We see that the sum of the interior angles of the quadrilateral equals the sum of the angles of these two triangles, i.e., $I_4 = 360$ deg.

Likewise, to find I_5, divide the pentagon into three triangles where $I_5 = 3(180) = 540$ deg. Notice that 180 is always multiplied by an integer that is 2 less than the number of sides, n, of the n-gon, in general:

$$I_n = (n-2)180\,\text{deg.} \tag{8.2}$$

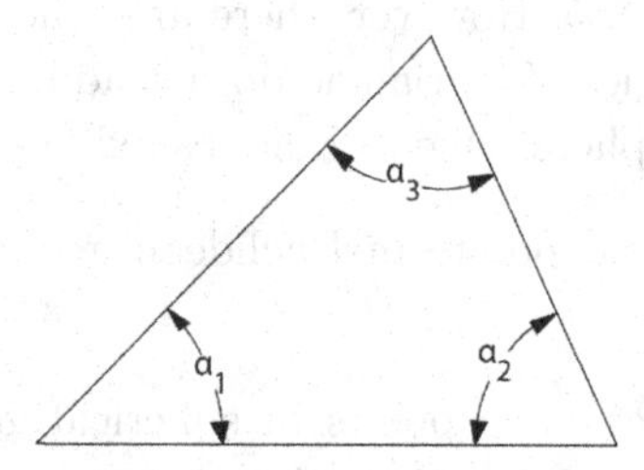

Fig. 8.2. The interior angles of a triangle.

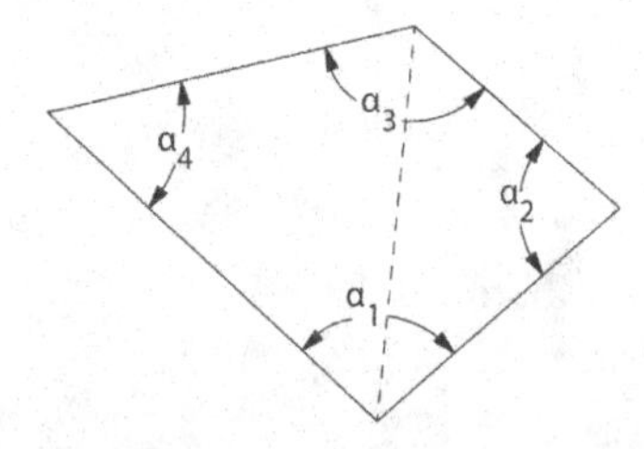

Fig. 8.3. The interior angles of a quadrilateral.

8.3. The Exterior Angles of a Triangle

Definition: Two angles are *supplementary* if they add up to 180 deg. as illustrated in Fig. 8.4 by angles α and β.

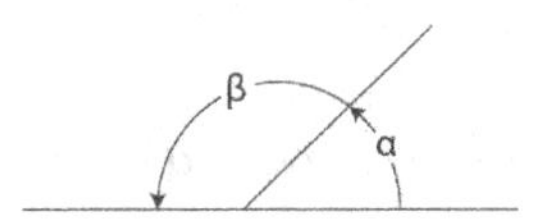

Fig. 8.4. Supplementary angles sum to 180 deg.

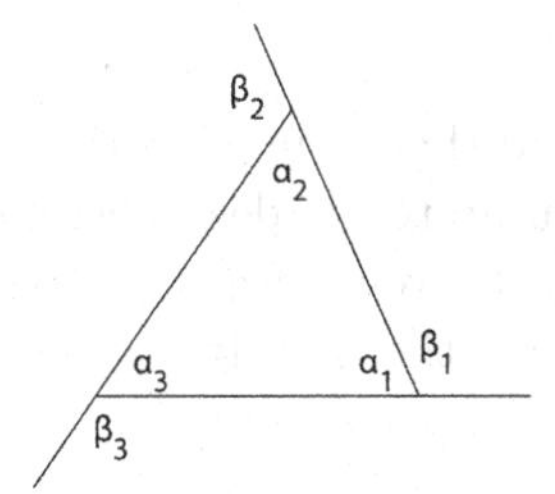

Fig. 8.5. Exterior angles of a triangle.

The angle supplementary to an internal angle of a triangle is referred to as an *exterior angle* denoted by the Greek letter, β. What is the sum of exterior angles of the triangle shown in Fig. 8.5? Since,

$$\beta_1 = 180 - \alpha_1, \quad \beta_2 = 180 - \alpha_2,$$

it follows that,

$$E_3 = \sum_{k=1}^{3} \beta_k = \sum_{k=1}^{3} (180 - \alpha_k) = 3(180) - \sum_{k=1}^{3} \alpha_k = 3(180) - 180 = 360. \tag{8.3}$$

Now consider the exterior angles of an n-gon. We simply generalize Eq. (8.3),

$$E_n = \sum_{k=1}^{n} \beta_k = \sum_{k=1}^{n} (180 - \alpha_k) = n(180) - \sum_{k=1}^{n} \alpha_k = n(180) - (n-2)180 = 360. \tag{8.4}$$

So we see that the exterior angles of any convex polygon always sum up to 360 deg. Do you see why that should be?

8.4. A Useful Relationship Between Interior and Exterior Angles

An exterior angle of a triangle (see Fig. 8.5) can be expressed in terms of the interior angles of the triangle. To do this, let

$$\beta_1 = 180 - \alpha_1 \tag{8.5}$$

where,

$$\alpha_1 = 180 - (\alpha_2 + \alpha_3). \tag{8.6}$$

Replacing Eq. (8.6) in Eq. (8.5),

$$\beta_1 = \alpha_2 + \alpha_3 \tag{8.7}$$

which is expressed as follows:

Theorem 8.1: *The exterior angle of a triangle is the sum of its alternate interior angles.*

It is important to understand that in proving this result, we made use of the fact that the angles of a triangle sum up to 180 deg. Therefore, as explained above, this is true only for Euclidean geometry. However, we can prove, without using the sum of the angles of a triangle being 180 deg., the slightly weaker result, important in the study of non-Euclidean geometries.

Theorem 8.2: *The exterior angle of a triangle is greater than either of its alternate interior angles.*

This will be proved in Appendix 8A.

Problem 1: As an exercise, generalize Theorem 8.1 to quadrilaterals and then to n-gons.

8.5. Interior Angles of a Regular Polygon

A polygon whose sides and angles are all equal is called a *regular polygon.* What is the interior angle θ of a regular polygon? We can answer that question in two ways:

Method 1: Simply take the sum of the interior angles of the polygon and divide it by the number of angles, i.e.,

$$\theta = \frac{I_n}{n} = \frac{(n-2)180}{n}. \tag{8.8}$$

Method 2: Since

$$\alpha_k = 180 - \beta_k,$$

$$I_n = \sum_{k=1}^{n}(180 - \beta_k) = 180n - \sum_{k=1}^{n}\beta_k = 180n - 360 = 180(n-2)$$

and

$$\theta = \frac{I_n}{n} = \frac{(n-2)180}{n}.$$

Problems:

2. For a regular pentagon:
   a. Compute its interior angle θ.
   b. If two non-intersecting chords are placed in a regular pentagon, it divides the pentagon into three isosceles triangles: one Triangle 1 and two Triangles 2. Find the angles of each triangle.

c. Taking the edge of the pentagon to be 1 unit, use the Law of Cosines to find the remaining edge lengths of Triangles 1 and 2. The answer should turn out to be 1.61803.... This number will be shown in Chap. 17 to have remarkable properties. It is called the *golden mean*.

3. For a regular hexagon:

   a. Compute its interior angle θ.
   b. Divide the hexagon into six triangles, meeting at the center of the hexagon, and show that these triangles are all equilateral.

4. For an octagon, (a) compute its interior angle; (b) divide the octagon into eight triangles meeting at the center of the octagon, and show that these triangles are 45, 45, 90-triangles.

8.6. Vertical Angles

When two lines intersect, the two pair of angles, α, β; γ, δ, at their intersection are called *vertical angles* as shown in Fig. 8.6. The following theorem pertains to vertical angles.

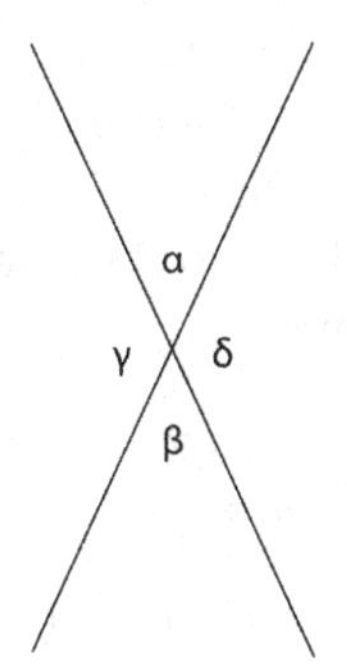

Fig. 8.6. Vertical angles.

Theorem 8.3: *Vertical angles are equal; i.e.,* $\alpha = \beta$ *and* $\gamma = \delta$.

Proof:

1. Consider Fig. 8.6.
2. The following angles are supplementary: α, δ; β, γ and α, γ.
3. $\alpha + \gamma = 180$.
4. $\beta + \gamma = 180$.
5. $\alpha + \delta = 180$.
6. Subtracting 4 from 3 yields: $\alpha = \beta$.
7. Subtracting 5 from 3 yields: $\gamma = \delta$.

QED

8.7. Radians and Degrees

The circumference of a circle is given by the formula, $C = 2\pi r$. If $r = 1$, then $C = 2\pi$. In terms of degrees, the total angle around a point is 360 deg. The circumference of a unit circle is taken as the measure of the angle of 360 deg. in terms of *radians*, i.e., 360 deg. $\equiv 2\pi$ rad. Therefore, the measure of 180 deg. in terms of radians is π rad. while the measure of 90 deg. is $\frac{\pi}{2}$ rad. We can now set up a ratio to determine the relationship between angles in degrees and radians,

$$\theta \text{ rad}/\theta \text{ deg.} = 2\pi \text{ rad.}/360 \text{ deg.},$$

or

$$\theta \text{ rad.} = \theta \text{ deg.} \left(\frac{\pi}{180}\right). \tag{9}$$

Remark: The length of an arc of a circle s of radius r subtending angle θ measured in radians is given by the formula, $s = r\theta$, which is a generalization of the formula for the circumference of a circle.

8.8. Angles Within a Circle

For a circle, a *central angle* is an angle with vertex at center O intercepting an arc, AB of the circle as shown in Fig. 8.7. An *inscribed angle* is an angle whose vertex O' lies on the circumference of the circle in the complement of arc AB and intercepts that arc. As you can see from Fig. 8.7, several such inscribed angles can intercept the same arc; each of these inscribed angles relates to the central angle by the following theorem (This is my favorite theorem of geometry):

Theorem 8.4: *An inscribed angle of a circle intercepting arc AB is one half of the central angle that intercepts the same arc.*

Proof: Try to prove this theorem for the diagram shown in Fig. 8.8. Hint: $\Delta O'OA$ and $\Delta BOO'$ are isosceles.

Remark: The same proof can be made to work for any other position of the inscribed angle.

Corollary: An inscribed angle intercepting a 180-deg. arc is a right angle.

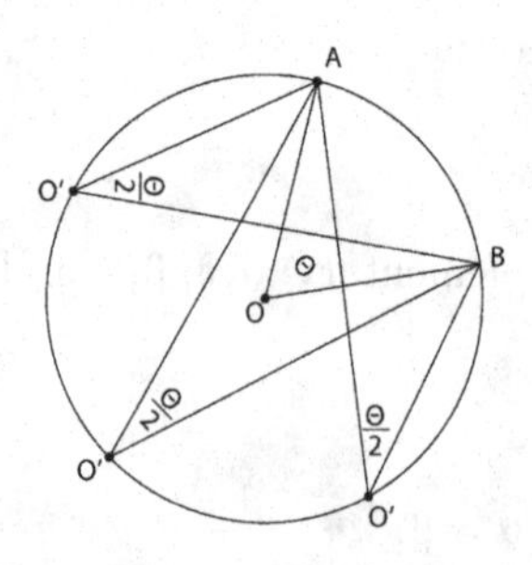

Fig. 8.7. Inscribed angles and a central angle of a circle intercepting the arc AB of a circle.

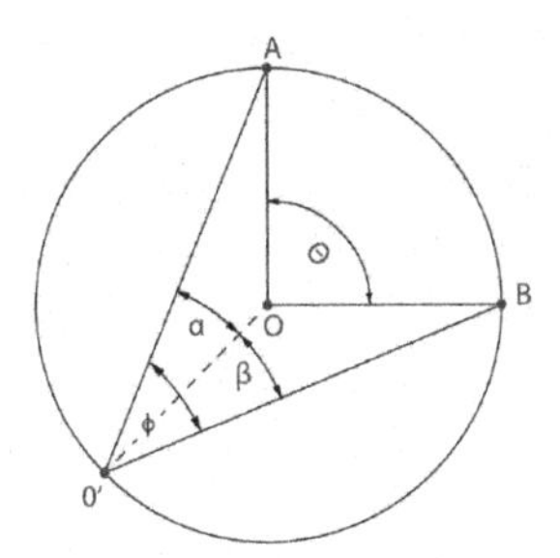

Fig. 8.8. Diagram illustrating the relationship between the central angle and the inscribed angle of a circle intercepting the same arc.

Remark: This corollary was used to locate position C of the right angle in Fig. 2.8b.

Appendix 8A. A Theorem of Importance to Non-Euclidean Geometry

Theorem 8.2: *The exterior angle of a triangle is greater than either of its alternate interior angles.*

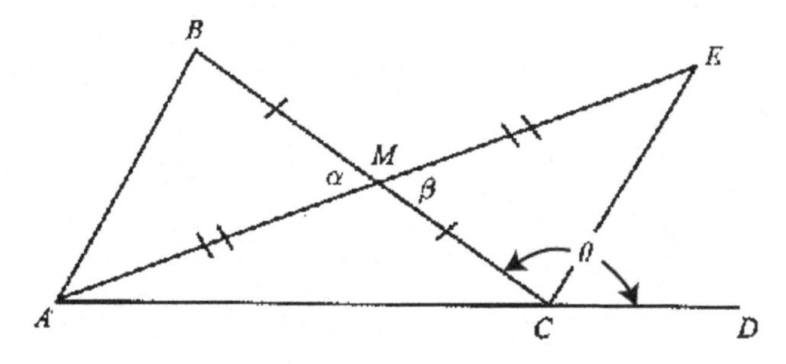

Fig. 8A.1. A diagram to prove Theorem 8.2.

Proof:

1. Consider ΔABC with exterior angle θ shown in Fig. 8A.1.
2. Let M be the midpoint of BC.
3. $BM = MC$.
4. Draw AM and extend it an equal distance to E.
5. $AM = ME$.
6. Draw EC splitting angle θ.
7. Angles: $\alpha = \beta$ since α and β are vertical angles.
8. $\Delta ABM \cong \Delta MCE$.
9. $\angle B = \angle MCE < \angle\theta$.
10. By a similar argument: $\angle A < \angle\theta$. QED

[illegible] the statement [illegible] the central [illegible] and the [illegible] of [illegible] is the same.

Remark: This [illegible] used to [illegible] the [illegible] in [illegible].

[illegible] A Theorem of Importance to Non-Euclidean Geometry

Theorem [illegible]. *The [illegible] angle of a [illegible] is [illegible] than [illegible] interior angles.*

Fig. [illegible] A [illegible]

Proof:

1. Consider $\triangle ABC$ [illegible] angle [illegible] shown [illegible]
2. [illegible] be the midpoint of AC.
3. $EM = MC$ [illegible]
4. Draw [illegible] equal distance to E [illegible]
5. [illegible]
6. Draw E [illegible] splitting angle C.
7. [illegible] vertical angles.
8. $\angle ECM = \angle MCB$ [illegible]
9. [illegible] $= M$ [illegible]
10. [illegible]

CHAPTER 9

VORONOI DOMAINS

9.1. Introduction

A map of proposed Cambridge, Massachusetts school districts is shown in Fig. 9.1. The black markers represent schools. The map is drawn so that each point in a school district is nearer to the school in that district than to any other school. Check a few points and show that this criterion holds for those points. School districts constructed in this way are called the *Voronoi* domains of the set of points corresponding to the schools. You will notice that in all cases but one, each vertex of the map is the meeting point of three districts (or Voronoi domains). Thus, each of the three schools from these three districts must lie on a circle whose center is at the vertex. (Why?) Why does it usually occur that a vertex of the map is surrounded by exactly three Voronoi domains? The one exception in Fig. 9.1 is surrounded by four Voronoi domains. Why is this such an unusual occurrence?

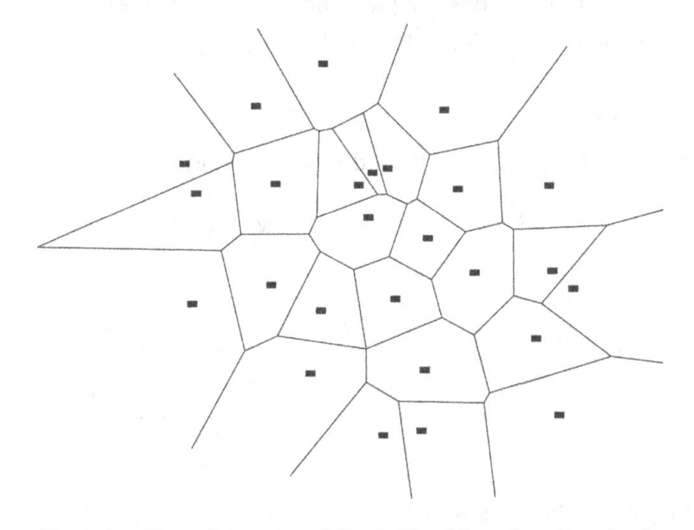

Fig. 9.1. Voronoi domains of Cambridge Massachusetts schools.

9.2. Construction of Voronoi Domains

We would like to find a way to construct the Voronoi domains of any set of points and thus be able to draw a map similar to Fig. 9.1.

9.2.1 Two points

First, consider the Voronoi domains corresponding to two points A and B shown in Fig. 9.2a. The Voronoi domains are clearly formed by bisecting the line segment AB with a perpendicular bisector. If you need a proof, here it is:

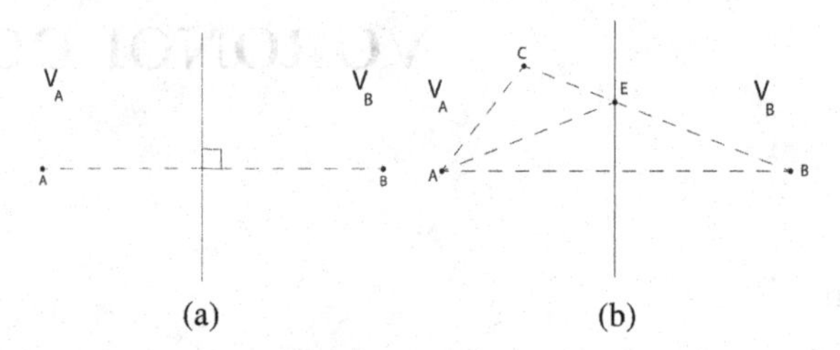

Fig. 9.2. (a) Voronoi domains of two points; (b) Proof that points in V_A are nearer to A than to B.

Given point C in Voronoi domain V_A, prove that C is nearer to A than to B.

Consider a point C in Voronoi domain V_A and point E where CB intersects the dividing line between V_A and V_B. (see Fig. 9.2b)

CA < CE + EA as a result of the triangle inequality
EA = EB (why?)
CA < CE + EB = CB

QED (the proof is complete)

The perpendicular bisector divides the plane into two half-planes in which A is in the left half-plane while B is in the right half-plane.

9.2.2 Three points

Now let us consider three points: A, B, and C shown in Fig. 9.3a.

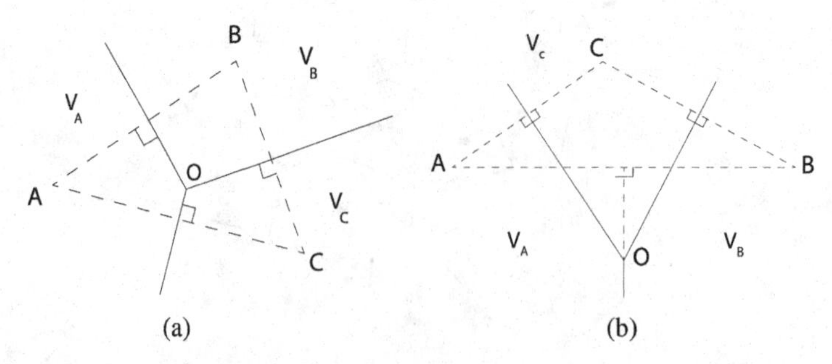

Fig. 9.3. (a) Voronoi domains of three points forming an acute triangle; (b) forming an obtuse triangle.

Clearly points on the perpendicular bisectors of BC, CA, and AB are equidistant from B and C, C and A, and A and B respectively. Also we now know, from Sec. 4.3b and the proof of Theorem 7.3, that the perpendicular bisectors of the sides of any triangle meet at a common point O. By similar reasoning as the proof in Sec. 9.2.1, the perpendicular bisectors pair to form inner envelopes that bound the Voronoi domains V_A, V_B, and V_C.

Remark 1: Figure 9.4 shows four lines l_1, l_2, l_3, l_4 marking the boundary of four half-planes enclosing some point P. Three of these half-planes with l_1, l_2, l_3, as boundaries form the inner envelope with l_4 outside of the envelope.

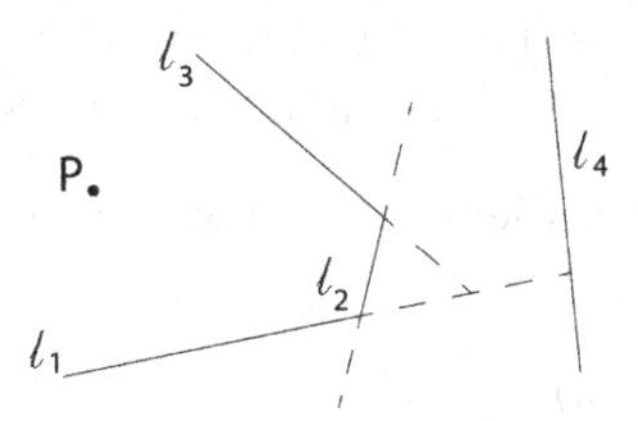

Fig. 9.4. The envelope of the boundary of four half-planes.

The geometry of these domains is shown in Fig. 9.3a where O plays an important role. Since all of the angles of ΔABC are acute, O lies within the triangle. Contrast this with the triangle in Fig. 9.3b which has an obtuse angle. Notice that O lies outside of the triangle.

All acute angled triangles have the meeting points of their perpendicular bisectors within the triangle whereas obtuse triangles have their meeting point outside. Right triangles have the meeting points on the hypotenuse.

In Figs. 9.3a the Voronoi domains are all convex. In fact, the inner envelope of half-planes forming the Voronoi domain is called the *convex hull* of the half-planes.

Definition: By a *convex domain* we mean that if I take any pair of points lying in the domain, the line drawn between them also lies in the domain. Figure 9.5a illustrates a convex domain while Fig. 9.5b is non-convex. We will find that for any number of points, the corresponding Voronoi domains will be convex.

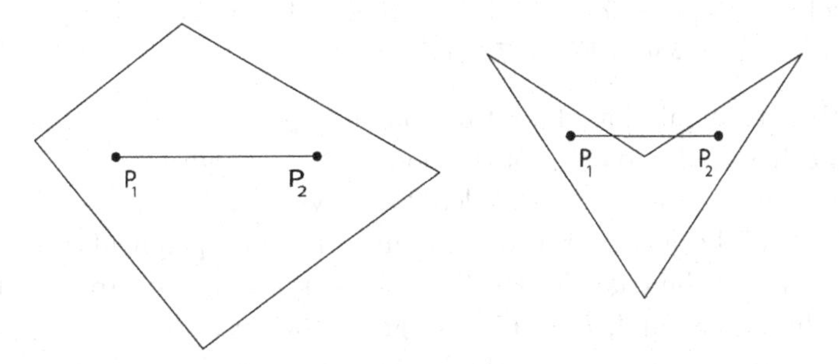

Fig. 9.5. (a) A convex domain; (b) a non-convex domain.

9.2.3 Four or more points

These ideas can be extended to Voronoi domains of four points A, B, C, and D as shown in Fig. 9.6 lying at the corners of a general quadrilateral (convex or non-convex) with all of its diagonals. How many perpendicular bisectors does this quadrilateral have? Clearly, it has as many perpendicular bisectors as there are distinct pairs of points. The number of distinct pairs of points are: AB, AC, AD, BC, BD, CD, i.e. six pairs. Contrast this with the number of point pairs for three points A, B, C: AB, AC, BC, i.e., three perpendicular bisectors as we found in Sec. 9.2.2. There is a general formula to predict the number of point pairs for n points, namely, $C(n, 2) = \frac{n!}{2!(n-2)!}$ where $C(n, 2)$ is the number of ways in which two objects can be drawn from a collection of n objects. $C(n, 2)$ is known as the combinatorial coefficient of n items taken two at a time. For example, $C(5, 2) = 10$ are the number of pairs of 5 points.

Remark 2: $C(n, k) = \frac{n!}{k!(n-k)!}$ is the number of ways that k objects can be drawn from a collection of n objects.

Remark 3: $n! = n(n-1)(n-2) \ldots 3.2.1.$

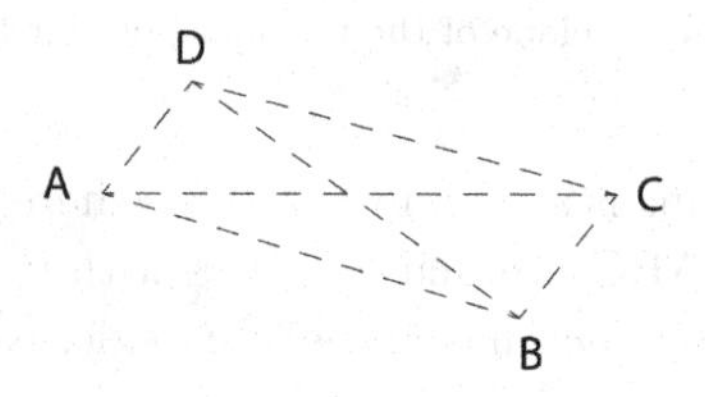

Fig. 9.6. Given four points there are 6 lines joining pairs of points.

The Voronoi domain of point P from a set of four points is again an inner envelope determined by the six half-planes containing P defined by the perpendicular bisectors of each of the six point pairs for the case of 4 points. Five points would have 10 half-planes to consider. Note that the number of half-planes proliferate rapidly as n gets large so that it is difficult to define the inner envelopes of these half-planes. We will look for a more manageable way to create Voronoi domains.

9.2.4 A simple way to construct the Voronoi domains of n points

Let us reconsider the case of four points in Fig. 9.6. There are two ways to triangulate this quadrilateral as shown in Fig. 9.7a and b.

Remark 4: By a triangulation of a set of points, I mean a set of triangles that do not intersect each other made up of all of the points of the quadrilateral as vertices.

The triangulation in Fig. 9.7a is called "bad" while the one in Fig. 9.7b is "good." Figure 9.7a is "bad" because the meeting points of the perpendicular bisectors of the triangles lie outside of the triangle (why?). This makes it difficult to draw the Voronoi domains. On the other hand, Fig. 9.7b is "good" because the meeting points all lie within the triangles. Because of this the Voronoi domains can be immediately drawn as in Sec. 9.2 for acute triangles as shown in Fig. 9.8.

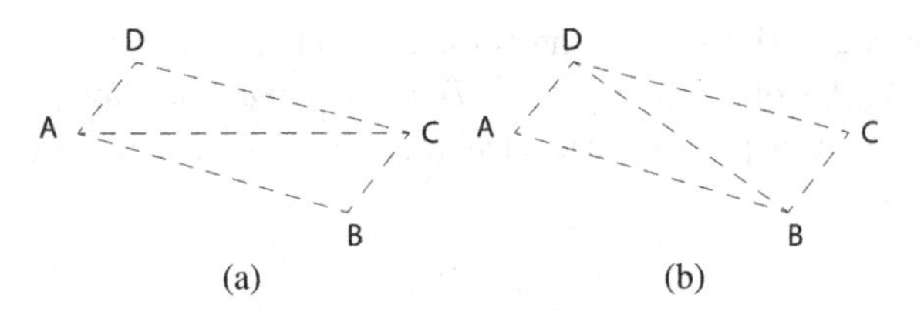

Fig. 9.7. Two triangulations of four points, (a) a "bad" triangulation (b) a "good" triangulation.

You just find the meeting point of the perpendicular bisectors within each acute angled triangle and connect one to another. Peripheral obtuse triangles can also be accommodated to this procedure.

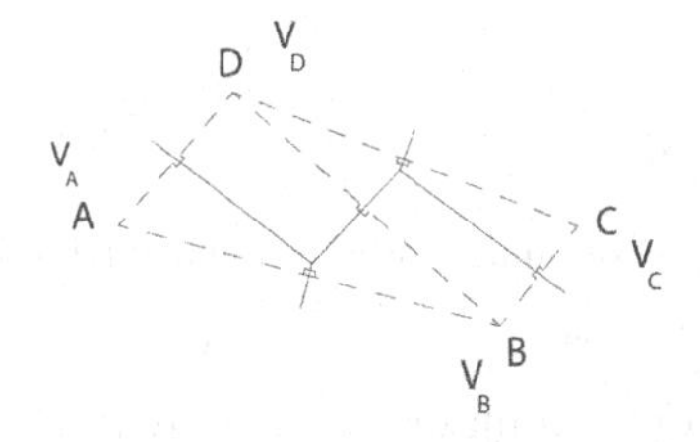

Fig. 9.8. Voronoi domains of four points with a good triangulation.

How can you distinguish, in general, a "good" triangulation from a "bad one"? We know that the meeting points of the perpendicular bisectors of a triangle lie at the center of its circumscribing circle. This circle is drawn in Fig. 9.9a for ΔABC of the triangulation in Fig. 9,7a whereas the circumscribing circle of ΔABD in the triangulation of Fig. 9.7b is shown in Fig. 9.9b. Do you notice that in the case of Fig. 9.9a, the fourth vertex D lies within the circle whereas in Fig. 9.9b the circle does not contain the fourth vertex C. So we now have a criteria to distinguish "good" from "bad" triangulation for any number of points.

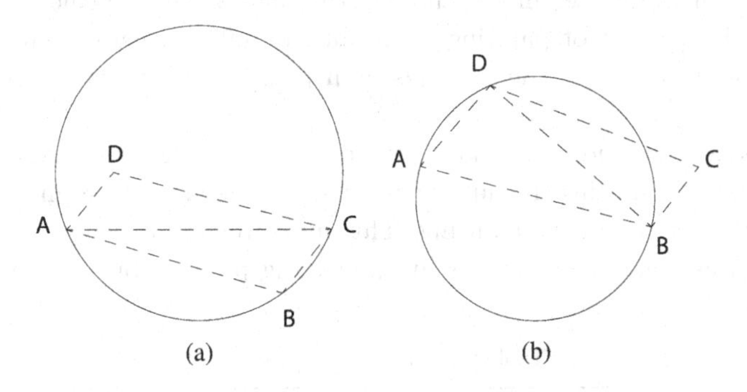

Fig. 9.9. Criteria for determining (a) a "bad" triangulation; (b) a "good" triangulation.

Criteria for distinguishing "good" from "bad" triangulations:

If none of the circumscribing circles of the triangulation contain points other than the triangles from which they were formed, then the triangulation is "good". Otherwise, it is bad. "Good" triangulations have the property that most of its triangles are acute. These "good" triangulations are known as *Delaunay triangulations.*

Figure 9.10 shows five points with a Delaunay triangulation along with its Voronoi domains.

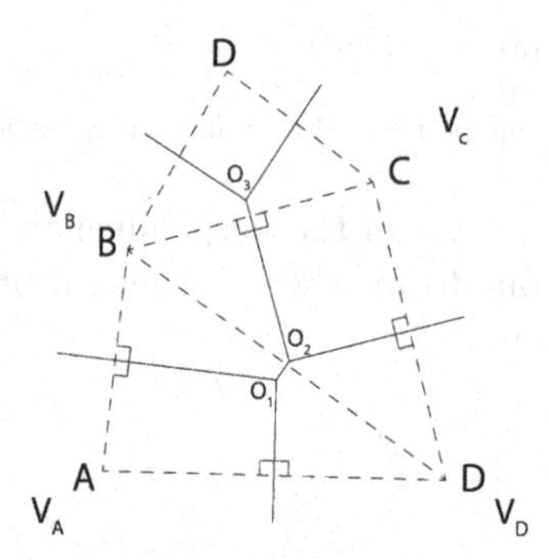

Fig. 9.10. Five points with a Delaunay triangulation.

Problem 1: Construct Voronoi domains for the diagrams of 2, 3, 4, and 5 points. Instead of using the constructions of Sec. 4.2.1 to construct perpendicular bisectors in these constructions, you may use a T-square or a right triangle and a straightedge.

Voronoi domains have applications that go beyond geometry.

9.3. Voronoi Domains and Pattern Recognition

A common problem in computer vision is to take a picture of an object and determine in which category it belongs. This is commonly called *pattern recognition.* As an example taken from [Mey], imagine a security robot that patrols various rooms in an art museum. A robot is not very good yet at recognizing what is in a painting, but they can readily estimate the lengths and heights of the paintings. The robot can be given a list of painting sizes, all carefully measured with great accuracy. For simplicity, let us say there are just four paintings in the museum, as shown in Table 9.1.

Now let us say the vision system is trying to determine the painting in the camera's field of view. We will call this the query painting, Q. The vision system determines the length and height of the query painting. This measurement of length and height via the camera and associated software is an interesting process, but it is also not exact

Table 9.1. Painting features.

Painting	Length (inches)	Height (inches)	Point in feature space
A	40	40	(40, 40)
B	60	50	(60, 50)
C	80	30	(80, 30)
D	70	20	(70, 20)

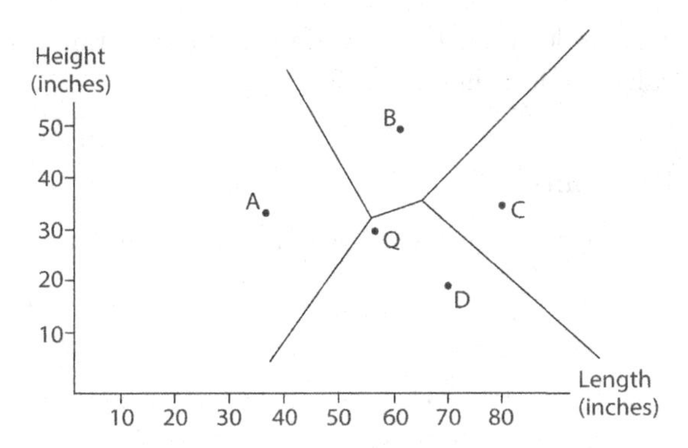

Fig. 9.11. Feature space for four paintings divided into Voronoi domains.

for a variety of reasons. The lighting may be bad, leading to an image that is hard to analyze, or light-colored painting frames may blend into the background color of the wall, making it hard for the robot to find the boundaries of the painting. For these and other reasons, the vision system's best guess about length and height of the query painting may be inaccurate and will not correspond precisely to any of the paintings which have been measured exactly. What the computer needs to do is to find the closest of the four exactly measured paintings to the query painting.

Let us turn this into a Voronoi problem. Begin by plotting a point on graph paper as in Fig. 9.11 for each of the four paintings in Table 9.1, the x-coordinate being the length, and y-coordinate, the height. Our graph paper with points interpreted as real or potential paintings is called the *feature space* for this pattern recognition problem. Next, draw the Voronoi domains of the four sites in feature space. Now suppose that the best guess for the query painting is the point $(58, 31)$. Plot this query point and see in which Voronoi domain it lies. We call this approach the *Voronoi diagram algorithm.*

It is instructive to compare this to an alternative approach we call the *distance formula algorithm.* In this approach, we do not use a Voronoi diagram. We simply calculate the distance from the query point to each of the four sites using the distance formula. For example, let us say the query painting is estimated to have length l_q and height h_q, so it would be represented by the point (l_q, h_q). To find the distance of the query painting to painting B, represented by $(60, 50)$, we use Eq. (3.1),

$$\text{Distance} = \sqrt{(l_q - 60)^2 + (h_q - 50)^2}.$$

Apply similar distance formulas to get the distance to paintings A, C and D. Pick the painting that is closest, and declare that to be the query painting.

The distance formula seems quicker if you are working by hand than the Voronoi diagram algorithm because drawing a Voronoi region is time consuming, and four times through the distance formula is not so bad with a hand calculator. On the other hand, suppose you had to do one hundred query points. Now the Voronoi approach starts to look better since you only have to draw the Voronoi diagram once. Then for every query point, you just have to plot the point and then visually observe in which Voronoi domain the query point is located. If you were using the distance formula algorithm you would have to calculate four hundred distances.

9.4. Additional Problems

2. Sketch the Voronoi diagrams for the lattice of points given in Fig. 9.12a and b.

(a) (b)

Fig. 9.12. (a) Square lattice; (b) Triangle lattice.

3. The vision system in Sec. 9.3 has estimated a query painting's length to be 65 inches and its height as 40 inches.
   (a) Use the Voronoi diagram algorithm and Fig. 9.11. Decide which painting this is most likely to be.
   (b) Answer the same question using the distance formula algorithm.
4. A query painting (see Sec. 9.3) has a length estimated as 85 inches. This is longer than the longest of the paintings (painting C) in Table 9.1. Can we conclude therefore that the closest of the sites is painting C? Or do we need to find the height and use either the Voronoi diagram algorithm or the distance formula algorithm? Explain your answer.
5. A washer used in machinery is a metal disk with a hole that is characterized by its outer diameter and the diameter of its hole. Table 9.2 shows washer types present in a bin. A computer vision system observes a washer and determines (inexactly) that the outer diameter is 13.5 millimeters and the diameter of its hole is 9.25 millimeters. Use the distance formula algorithm to determine which type it is most likely to be. Then use the Voronoi diagram algorithm. Do you get the same results? Do your work on graph paper.

Table 9.2. Washer features.

Type	Diameter of Hole (mm)	Outer Diameter (mm)
A	5	10
B	4	16
C	10	15
D	7	13
E	9	11

6. This problem does not use Voronoi domains or the distance formula but it uses a method in the same spirit. Fire engines cannot go in a straight line to a fire — they need to follow streets that usually have a rectangular pattern, as illustrated in Fig. 9.13a. Shade the streets, or parts of streets that are closer to S_1 than S_2. Two such streets are already shaded as an example. Here distance is measured along the street grid so that the distance from the location of the fire to S_1 is two blocks, not $\sqrt{1^2 + 1^2}$ as the crow flies. The distance between two coordinates of the street grid between points (x_1, y_1) and (x_2, y_2) is given by the formula: $|x_2 - x_1| + |y_2 - y_1|$.

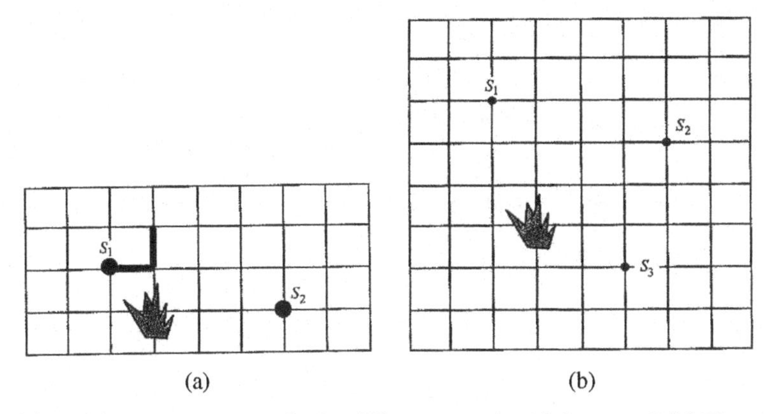

Fig. 9.13. (a) Two fire stations sites and a fire. What streets should they serve? (b) Three fire stations and a fire. What streets should they serve?

7. Draw the Voronoi diagram based on distance along the street grid rather than a straight line for the three fire stations in Fig. 9.13b. This problem should be done in the spirit of the previous exercise. Use colored pencils to color line segments in different colors depending on whether they are nearer to $\mathrm{V}(S_1), \mathrm{V}(S_2)$, or $\mathrm{V}(S_3)$. Ignore points that are not on streets. The end result will be a color-coded system of streets that enable the fire stations to visually see their fire-fighting domains.

CHAPTER 10

PARALLEL LINES

10.1. Introduction

In Euclidean geometry, either two lines intersect or they are parallel, meaning that they do not intersect no matter how far they are extended. One can always find the angle at which the lines intersect. For parallel lines, as you will see, we can say, in some limiting sense, that they intersect at 0 deg.

To get a better feel for the nature of parallel lines, try this experiment. Select a tall tree in your neighborhood, and point your finger towards it. Now walk 100 ft. at right angles to your original direction to the tree, and point again. You will notice that the direction of your pointing finger has changed. Now point to the sun. Walk 100 ft., and point again. Do you notice that your pointing direction has not changed? Why? Actually the distance to the sun is far greater than 100 ft. so that the distance to the sun is effectively infinite by comparison to the tree. As Fig. 10.1 shows, the further a point moves from the 100 ft. baseline, the smaller is the angle that the point intercepts. In the limit, this angle becomes effectively 0 deg. Therefore, all of the light rays from the sun or the moon reach points on Earth effectively in a parallel direction (see Fig. 10.2).

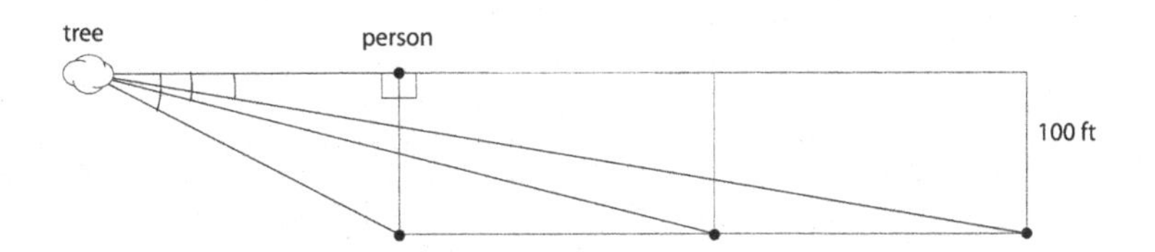

Fig. 10.1. Angle between parallel lines approaches 0 deg.

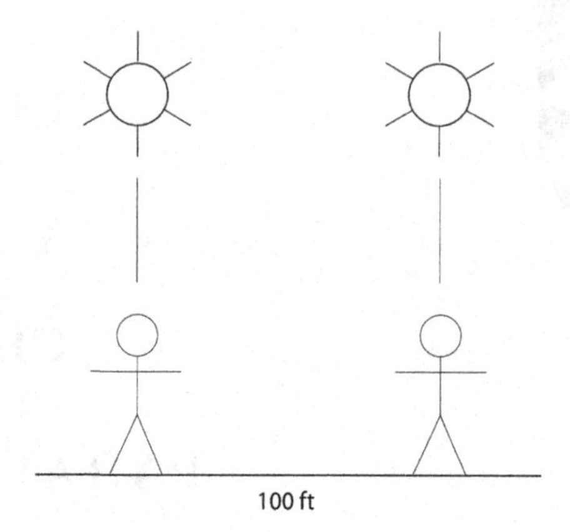

Fig. 10.2. The sun's rays incident to the Earth are parallel.

10.2. Euclid's Parallel Axiom

The fifth Axiom of Euclid states:

Axiom 5: Given a line l and a point P, only one line can be drawn through P parallel to l.

Despite the fact that this axiom appears to be self-evident, it disturbed mathematicians for centuries after it was first stated by Euclid. It seemed to stand alone in that many of Euclid's theorems could be proven without the use of this axiom. In fact, mathematicians tried to prove Axiom 5 from the other axioms, but they failed in their attempts. Finally, in the early 19th century, two mathematicians, Bolyai and Lobeshevsky, independently discovered other non-Euclidean geometries that negated Axiom 5. In one model of such a non-Euclidean geometry known as the *Poincare model*, the points at infinity are represented by a circle. Given two points within the circle, a line is defined to be the unique arc of a circle that contains the two points in the circle and meets the circle at infinity at right angles as shown in Fig. 10.3.

A construction with compass and straightedge to construct the unique arc of a circle meeting the circle of infinity at right angles is presented in Appendix 10A. The

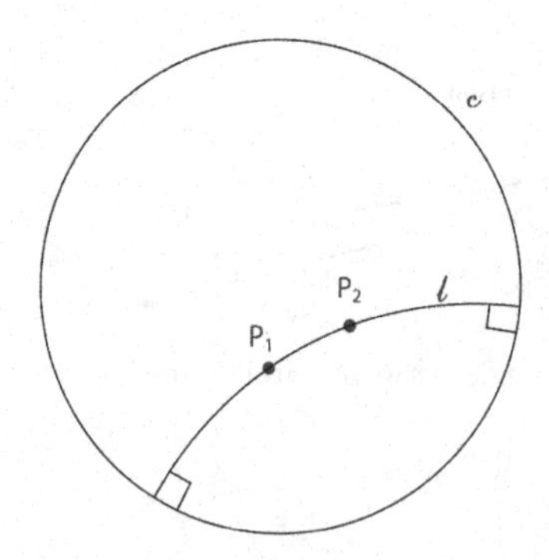

Fig. 10.3. Two points define a unique line in Poincare geometry.

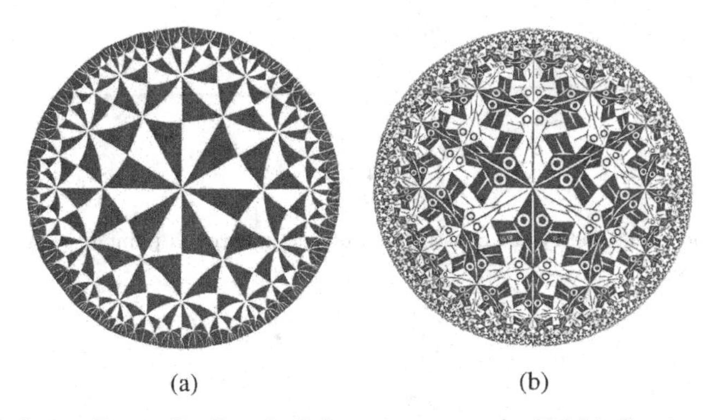

(a) (b)

Fig. 10.4. (a) A design illustrating lines in Poincare geometry by H.S.M. Coxeter; (b) Circle limit I by M.C. Escher.

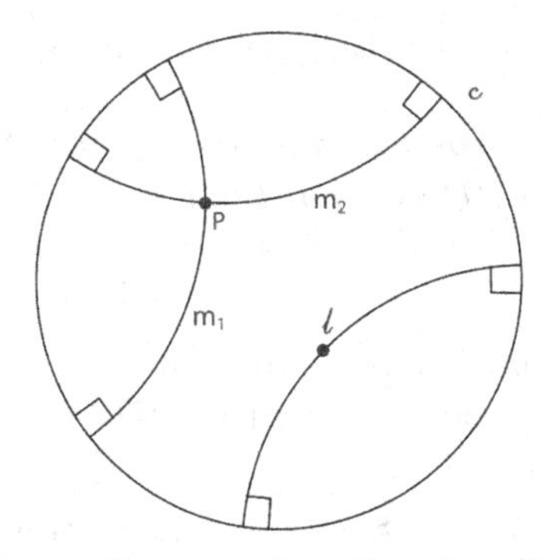

Fig. 10.5. An illustration in Poincare Geometry of two lines through point P parallel to a given line l.

great twentieth century geometer, H. S. M. Coxeter, illustrates this non-Euclidean geometry in Fig. 10.4a. Later, the twentieth century artist M. C. Escher established a communication with Coxeter and rendered one of his great lithographs entitled "Circle Limit 1" depicting this geometry illustrated in Fig. 10.4b. In this geometry, it is easy to find a contradiction to Axiom 5 as we have done in Fig. 10.5 where "lines" m_1 and m_2 through P are parallel to l, i.e., they do not intersect.

10.3. Parallel Lines in Euclidean Geometry

Returning to Euclidean geometry, given a line l and a point P not on l, we can construct the unique line through P parallel to l as follows (see Fig. 10.6):

a. Construct a line through P perpendicular to l with D where the line intersects l.
b. Through P, construct another line perpendicular to DP.
c. This line must be parallel to l.

Although this construction may appear obvious, it requires a proof, and this proof indicates that this theorem is limited to Euclidean geometry.

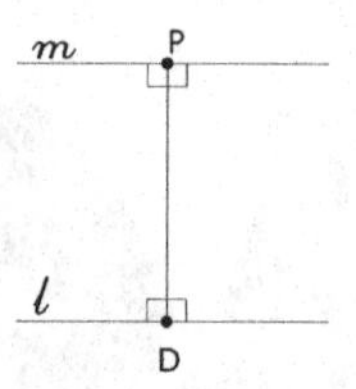

Fig. 10.6. In Euclidean geometry the construction of line m through point P parallel to given line l.

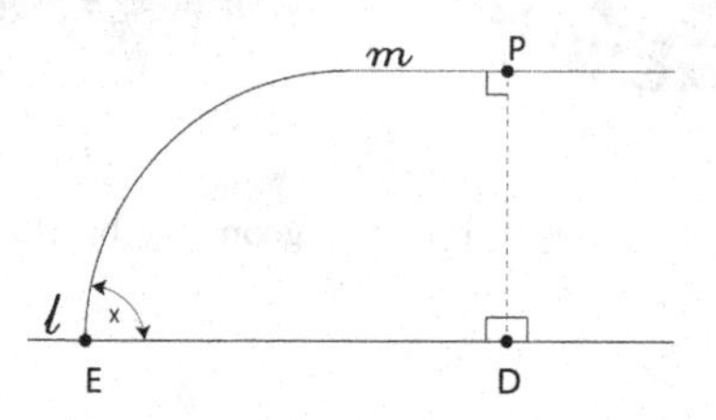

Fig. 10.7. Proof that two parallel lines have no point of intersection in Euclidean geometry.

Theorem 10.1: *Given line l, point P, and line segment PD perpendicular to l, line m through P perpendicular to PD is parallel to l.*

Proof.

1. Consider line l, point P, and PD perpendicular to l (see Fig. 10.7).
2. Construct line m through P perpendicular to PD.
3. Assume that m and l are not parallel.
4. m then meets l at some point E shown in Fig. 10.7, forming triangle PED with $PED = x > 0$.
5. $90 = 90 + x$ or $x = 0\,\text{deg.}$ follows from Theorem 8.1.
6. Contradiction.
7. Therefore m and l must be parallel.

QED

Remark: As we showed in Theorem 8.1, Step 5 of this proof requires that the angles of a triangle sum up to 180 deg. which is not true for non-Euclidean geometries.

Remark: This approach to proof is known as "proof by contradiction." We negate the conclusion and assume that the hypothesis is true and show that this leads to a contradiction which implies that the conclusion must be valid.

10.4. Conditions for Two Lines Being Parallel

Next, consider two parallel lines l and m, cut by an oblique line called a *transversal* (see Fig. 10.8) with the two angles α, β referred to as *alternate interior angles.* I can now state the following theorem which expresses the most fundamental property of parallel lines:

Theorem 10.2: *Given two parallel lines, l and m cut by a transversal AE (see Fig. 10.9), the alternate interior angles are equal.*

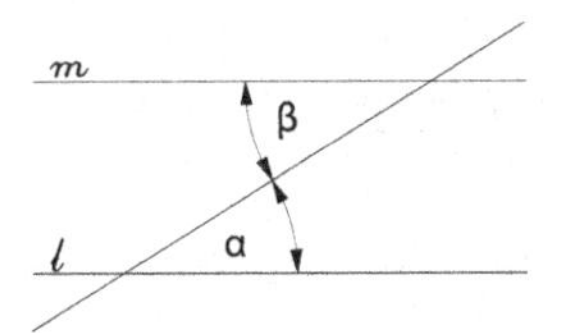

Fig. 10.8. Two parallel lines cut by a transversal revealing alternate interior angles.

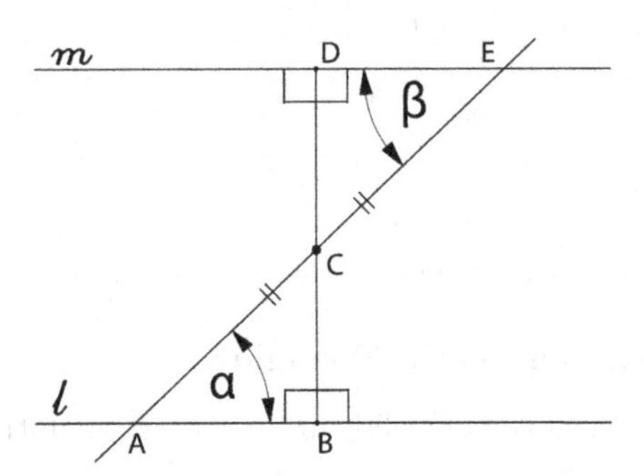

Fig. 10.9. Proof that the alternate interior angles are equal.

Proof.

1. Consider lines l, m and transversal AE.
2. Let C be the midpoint of transversal AE.
3. $AC = CE$.
4. Drop a line from C perpendicular to l, meeting l at point B.
5. Extend CB to m, intersecting m at point D.
6. $\angle ABC = 90\,\text{deg}$.
7. $\angle EDC = 90\,\text{deg}$.
8. $\angle BCA = \angle DCE$.
9. $\Delta ABC \cong \Delta EDC$.
10. $\angle\alpha = \angle\beta$.

QED

Remark: Step 7 requires some justification. By dropping a perpendicular from D to l, it follows that $\angle ABC = 90\,\text{deg}$. But how do we know that $\angle EDC = 90\,\text{deg}$.? It was given that m is parallel to l, and from Theorem 10.1 a line perpendicular to BD must be parallel to l. Since, by Axiom 5 of Euclidean geometry, only one line through D can be parallel to l, that line must also be perpendicular to BD, and so it must follow that $\angle ECD = 90\,\text{deg}$. Now, supply the other justifications to the steps in the proof of Theorem 10.2.

In Fig. 10.10, angles α, β are called *complementary angles.* I can now state another useful theorem about parallel lines:

Theorem 10.3: *Given two parallel lines cut by a transversal, complementary angles are equal.*

The proof is left to the student.

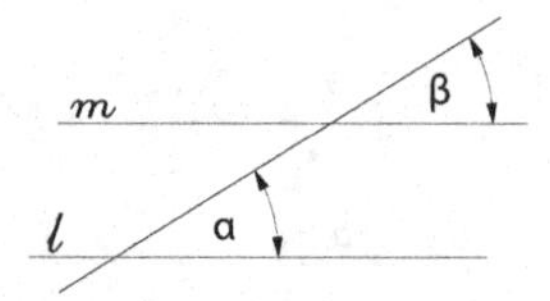

Fig. 10.10. When two parallel lines are cut by a transversal, complementary angles are equal.

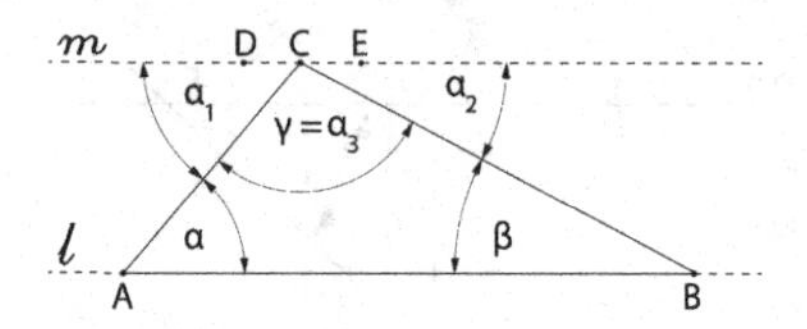

Fig. 10.11. The sum of the interior angles of a triangle equal 180 deg.

10.5. The Sum of the Angles of a Triangle

We are now able to prove, within Euclidean geometry, that the angles of a triangle sum to 180 deg.

Theorem 10.4: *The angles of a triangle sum to 180 deg.*

Proof.

1. Consider ΔABC with angles α, β, γ (see Fig. 10.11).
2. Extend AB to line l.
3. Through C, draw the unique line m parallel to l.
4. Mark off points D and E on line m.
5. Let $\angle ACD = \alpha_1$ and $\angle ECB = \alpha_2$.
6. $\alpha = \alpha_1$, $\beta = \alpha_2$.
7. Let $\angle BCA = \alpha_3$.
8. $\alpha + \beta + \gamma = \alpha_1 + \alpha_2 + \alpha_3 = 180\,\text{deg}$.

QED

Remark: Step 6 follows from Theorem 10.2. Step 3 shows why this theorem is limited to Euclidean geometry. The student can fill in the remaining justifications.

You can use Theorem 10.2 to prove Theorems 10.5 and 10.6.

Theorem 10.5: *The diagonals of a parallelogram bisect each other* (*see Fig.* 10.11a).

Theorem 10.6: *If a four-sided figure is a rhombus* (*parallelogram with all sides equal*), *then the diagonals bisect each other and are perpendicular* (*see Fig.* 10.11b).

Problems 1 and 2: Prove Theorems 10.5 and 10.6.

10.6. An Application of Parallel Lines to Finding the Circumference of the Earth

Eratosthenes (280–195 BC), a great scientist of antiquity, made use of the parallel light rays from the sun (see Sec. 10.1) to accurately measure the circumference C of Earth.

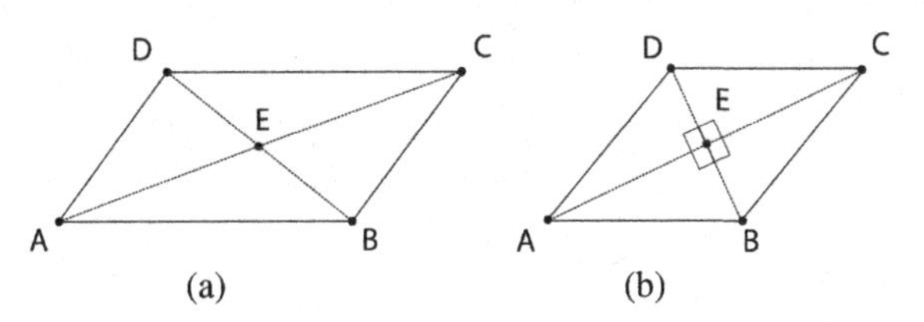

Fig. 10.12. (a) Diagonals of a parallelogram bisect each other; (b) Diagonals of a parallelogram with all equal sides bisect each other and are perpendicular.

What he did was to measure the distance from the city of Styrene, S, to Alexandria, A, by utilizing a caravan route with camels to walk off the distance between these cities. He counted the steps of the camel and then measured the length of a camel step. In this way, he found the distance to be 500 mi., equivalent to arc AS subtending an unknown angle θ as shown in Fig. 10.13a.

Eratosthenes knew that on a certain day of the year, the sun's rays penetrated to the bottom of a well in Styrene. Therefore, on that day, the rays penetrated towards the center of Earth. At Alexandria, Eratosthenes erected a pole and measured the shadow of the pole as shown in Fig. 10.13b. In this way, he was able to measure $\angle\phi$. But since light rays at Styrene and Alexandria were effectively parallel, by making use of Theorem 10.2, $\angle\phi = \angle\theta$. Eratosthenes found that $\angle\phi = 7.2$ deg. $= \angle\theta$. He then set up the proportion,

$$\frac{500}{7.2} = \frac{C}{360}$$

or,

$$C = (500)\frac{360}{7.2}.$$

Since $\frac{360}{7.2} = 50$, it followed that the circumference,

$$C = 500 \times 50 = 25{,}000\ mi.,$$

which is close to the modern value of 24,860.

Problem 3: We saw in Sec. 8.7 that

$$S = r\theta \tag{10.1}$$

where S is the arc length on a circle and θ is the central angle. It is important that θ is measured in units of radians (see Sec. 8.6), i.e.,

$$\theta_{rad} = \frac{\pi}{180}\theta_{\text{deg}}. \tag{10.2}$$

Given that $S = 500$ mi. between Styrene and Alexandria and $\theta = 7.2$ deg., use Eqs. (10.1) and (10.2) to find the radius of Earth. Check your result with the actual value of 3693 mi.

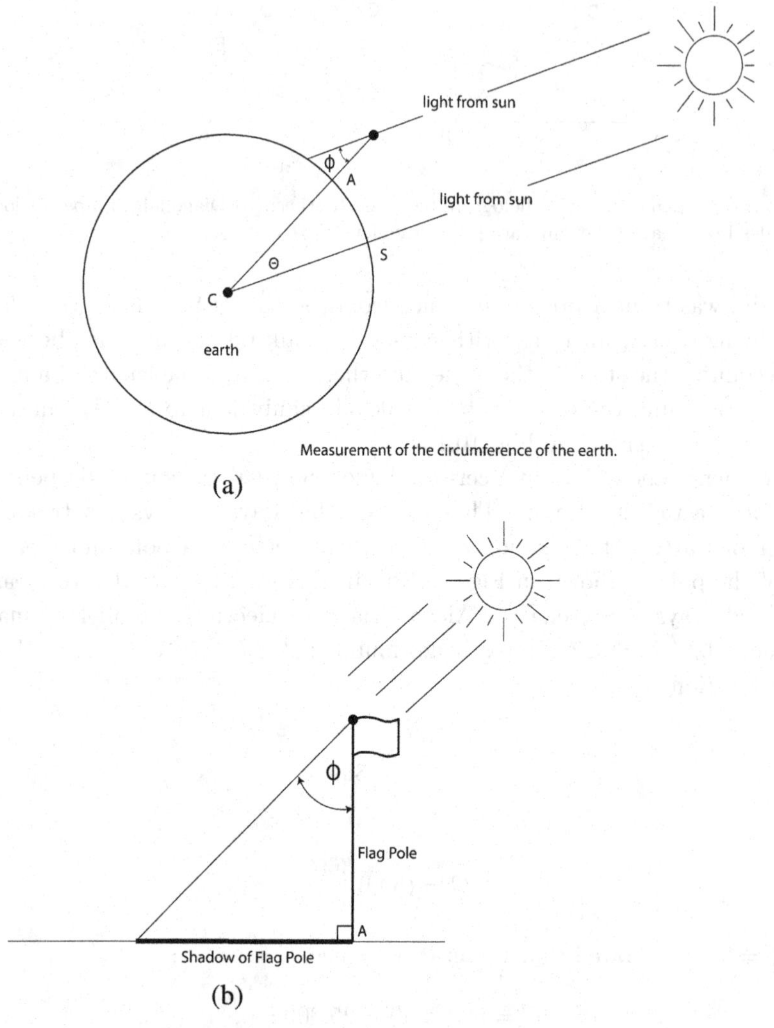

Fig. 10.13. (a) Eratosthenes' measurement of the circumference of the Earth; (b) detail of the shadow cast by the flagpole at Alexandria.

Appendix 10A. The Geometry of Lines in the Poincare Plane

The Poincare plane provides a model for non-Euclidean geometry. It consists of a circle C representing all the points at infinity. Inside the circle lie the finite points of the plane. A line in the finite plane is defined as the arc of a circle that meets circle C at right angles. Through any point P in the finite plane (within the circle), many lines (arcs of circles) can be drawn. However, given two points P_1 and P_2 in the finite plane, there is a unique arc of a circle (line) that meets C at right angles.

We show how to construct, using compass and straightedge, many "lines" through point P. Consider circle C to have center O and unit radius as shown in Fig. 10A.1a. Let point P lie within circle C distant from center O by OP. On the extension of line OP, construct point P' such that $OP' = 1/OP$. Construct the perpendicular bisector

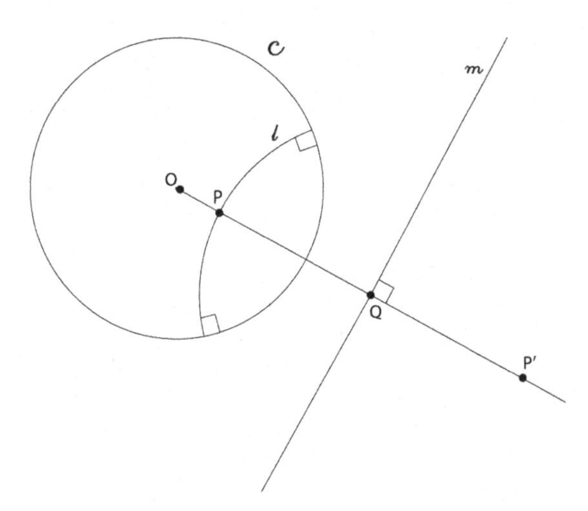

Fig. 10A.1a. Construction of a "line" in the Poincare geometry.

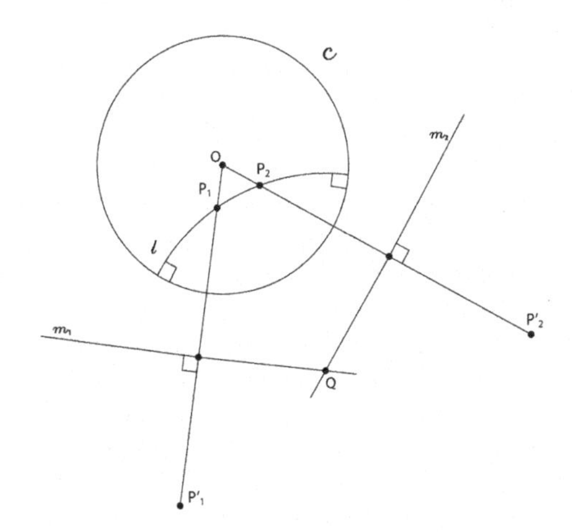

Fig. 10A.1b. Two points define a unique line in Poincare geometry.

m of PP'. I claim that each point on m is the center of a circle that goes through P and meets C at right angles. A "line" with center at Q is shown in Fig. 10A.1a. Therefore, each of these arcs represents a Poincare line through P.

a. Construct the unique "line" l through P_1 and P_2.

Using the result of (a), construct perpendicular bisector m_1 of P_1P_1' and m_2 of P_2P_2'. The unique center Q of a circle that goes through P_1 and P_2 is located where m_1 and m_2 intersect (see Fig. 10A.1b) and meets C at right angles. An arc of this circle is the unique "line" through P_1 and P_2 in Poincare geometry.

Remark: Points P and P' with respect to C, where $OP' = 1/OP$, is known as inversion in the circle (see Appendix A13).

CHAPTER 11

BRACING A FRAMEWORK

11.1. Introduction

We are going to study the bracing of a framework grid with m rows and n columns, i.e., an $m \times n$ grid by using a set of crossbars [kap 1].

You will need some cardboard, scissors and some nails or paper clips.

If the edges of the grid measure $2\frac{1}{2}$ inches, then the crossbars should measure $3\frac{1}{2}$ inches from hole to hole as shown in Fig. 11.1. The elements of the framework and the crossbars should be about $\frac{1}{4}$ inch wide. You will need 24 edges and 7 crossbars.

11.2. Frameworks

a. Construct a 3×3 framework.
b. To make the framework rigid, every square need not be braced. Determine, by testing it on your framework, the minimum number of crossbars needed to make the entire structure rigid.
c. Sketch several configurations and determine whether they are rigid by testing it on your framework. At least one of the configurations should be rigid with the minimum number of crossbars (*minimally braced*); one should be rigid with more crossbars than the minimal (*overbraced*), and one should not be rigid (*underbraced*).

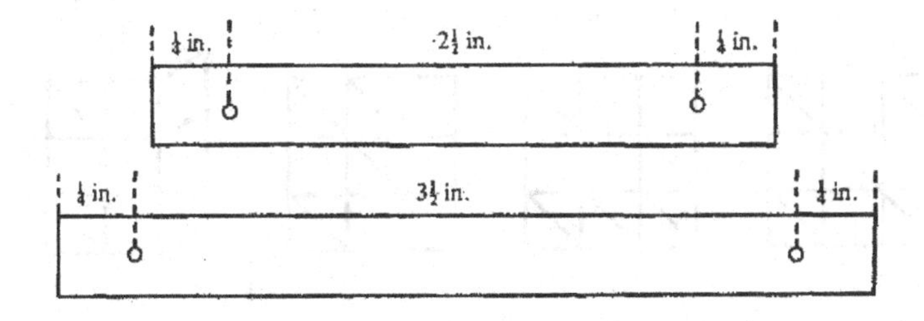

Fig. 11.1. Cardboard strips for a bracing experiment.

d. Use the method in Sec. 11.3 to prove whether your framework is rigid or not rigid. The method is based on the following lemma.

Lemma: *In any distorted grid, all the elements of a row (column) are parallel (see Fig.* 11.2).

e. Figure 11.3 shows nine 3×3 grids. Use your 3×3 framework to show that Fig. 11.3e is rigid while Fig. 11.3g is not. Then use the method in the next section to prove this.

Remark: A lemma is a theorem that is used to prove another theorem.

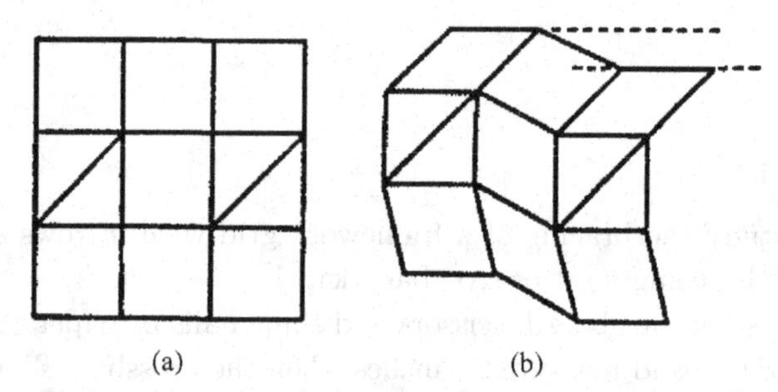

Fig. 11.2. (a) Horizontal and vertical edges of a framework; (b) the edges in each row or column remain parallel when they distort.

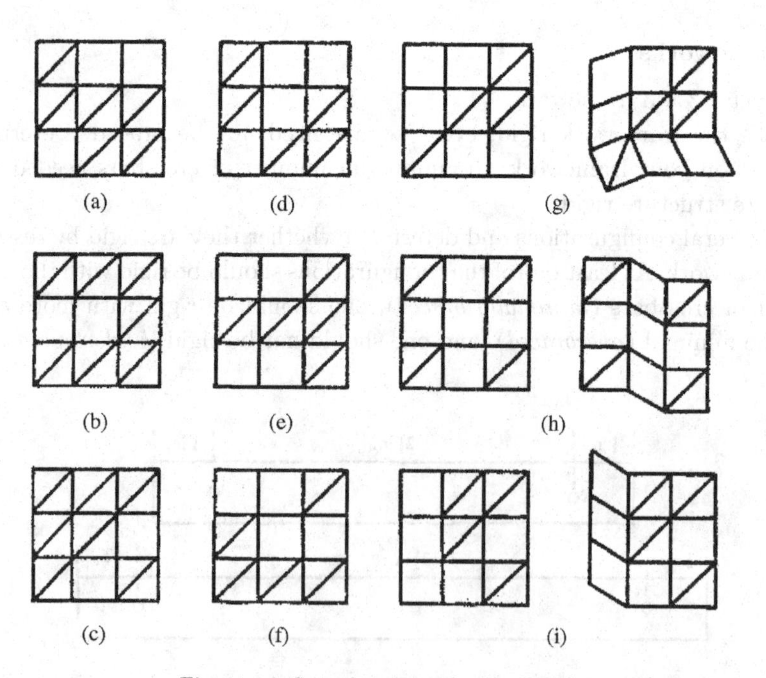

Fig. 11.3. Some bracings of a 3×3 grid.

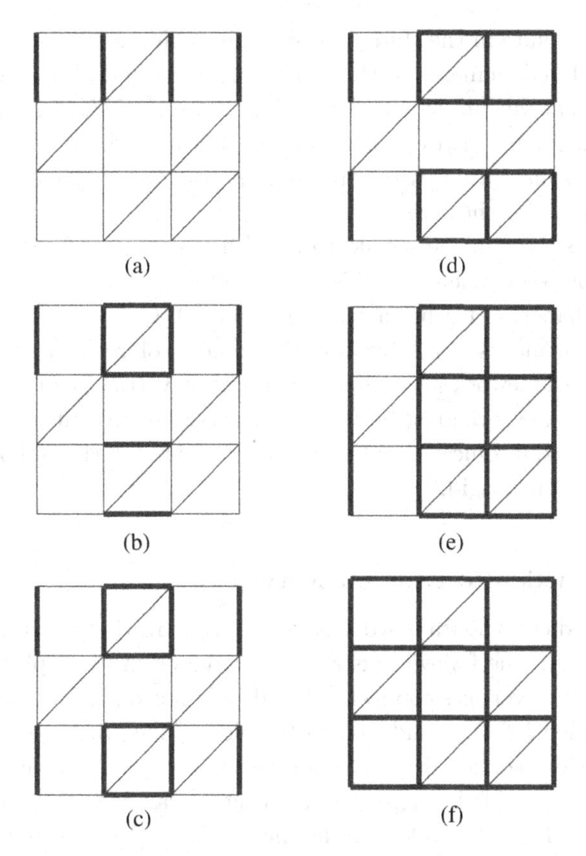

Fig. 11.4. (a)–(f) Steps of an algorithm to determine the rigidity of a framework.

11.3. An Algorithm to Determine Rigidity of a Framework

To determine whether a framework is rigid or not, we use the following procedure illustrated for the framework in Fig. 11.4:

a. You can begin, in all generality, by making the upper left-hand element of the framework vertical (see Fig. 11.4a). Since, by the lemma in flexing, edges of any row or column remain parallel, all of the upright elements of row 1 are also vertical. Color these vertical elements in red (bold lines in Fig. 11.4).
b. Notice in the top row that the middle square is braced, so it will remain a square and, therefore, the remaining two elements of this square must be horizontal (see Fig. 11.4b). Therefore, when flexing all of the parallel elements in the second column, they must also be horizontal. Color these elements in red.
c. Next, notice that the square in the bottom row of column 2 is braced. Therefore, the remaining elements of this square must be vertical, and, when flexing, all of the upright elements in the third row must also be vertical (see Fig. 11.4c). Color these in red.

d. Notice that the square in the third row and third column is braced. As a result, the remaining uncolored elements of this square must be horizontal. And so also must the other parallel elements of the third column (see Fig. 11.4d). Color these in red.
e. Notice that the square in row 2 and column 3 is braced. As a result, the upright elements of this square are vertical, and so must the other upright elements of row 2 (see Fig. 11.4e). Color these red.
f. Notice that the square in the second row and first column is braced. Therefore, the remaining uncolored elements of this square must be horizontal, and so must the other parallel elements of column 1 (see Fig. 11.4f). Color these red.
g. Since all of the elements of the framework are now colored red, this means that all elements of the framework are either horizontal or vertical and remain as squares. Therefore, this framework must be rigid; it does not flex from its original position. In another framework, if some elements are not colored red at the end of this procedure, that framework is not rigid.

11.4. An Introduction to Graph Theory

Another way to predict if the framework is rigid or not, overbraced or minimally braced involves graph theory. The following is a brief introduction to graph theory:

A *graph* is a set of vertices connected by edges. Figure 11.5a illustrates a graph. It has 4 vertices (labeled $1, 2, 3, 4$) and 4 edges (labeled a, b, c, d). Notice that edges d and a intersect but the intersection is not considered to be a vertex. In fact, we can imagine the vertices to be pins and the edges to be strings. I can remove a pin and replace it at a new position in Fig. 11.5b. As long as the vertices are connected by edges, in the same way, the graph is considered to be identical. I can also consider a *path* within the graph which is a sequence of vertices and edges, e,g., $1a2b3c1$ is a path. Notice that this path comes back to where it started. Therefore, we say that this graph has a *cycle*.

Next consider the graph in Fig. 11.6. It has 6 vertices and 5 edges. But notice that it has no cycles. A graph with no cycles is called a *tree*. It is easy to see that a tree always has one more vertex than it has for edges.

Now consider the graph in Fig. 11.7. Notice that it comes in two pieces and some vertices cannot be reached from others, e.g., there is no path between vertices 1 and 5.

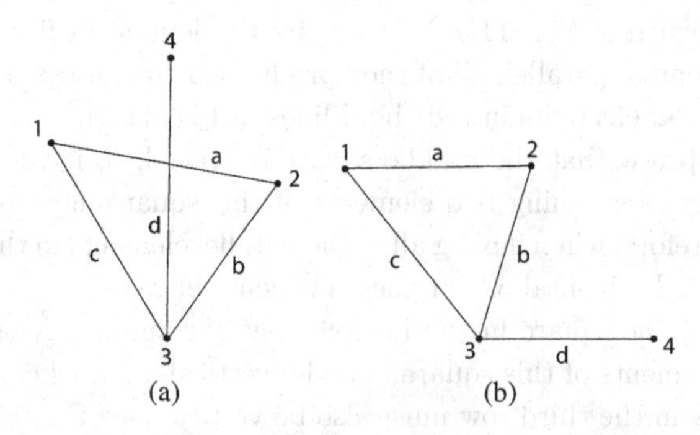

Fig. 11.5. (a) A graph; (b) the same graph with no crossovers.

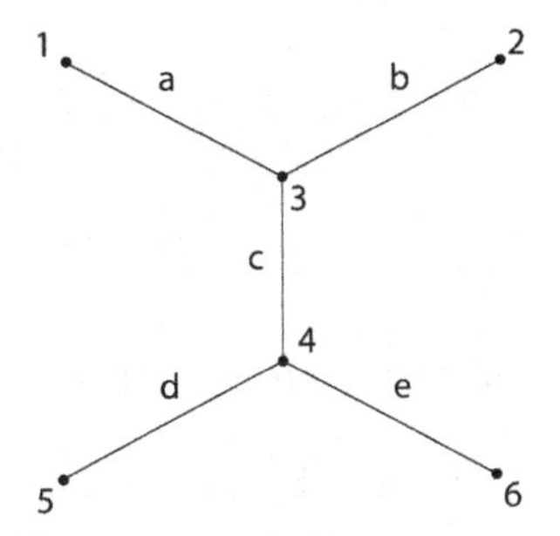

Fig. 11.6. A tree graph.

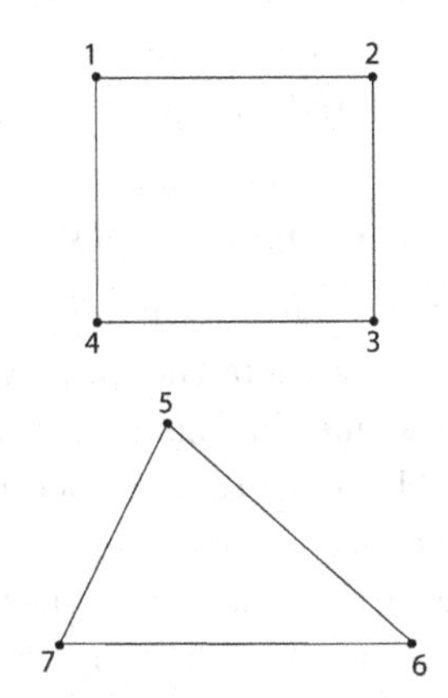

Fig. 11.7. A graph that is not connected.

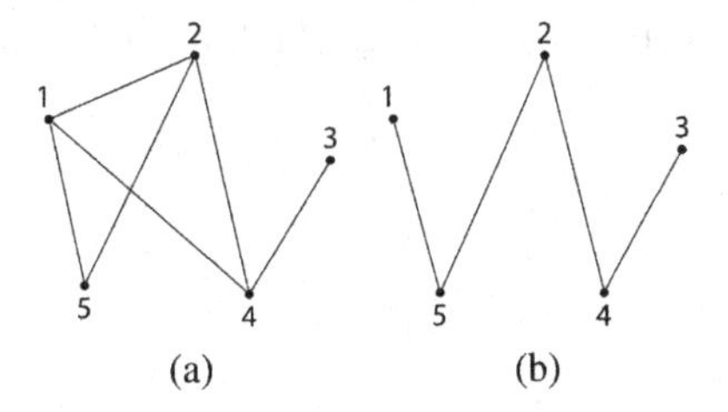

Fig. 11.8. (a) A graph G_1; (b) a subgraph G_2.

Such a graph is said to be *disconnected*. If there is a path between any two vertices, the graph is said to be *connected*.

Notice that graphs G_1 and G_2 in Figs. 11.8a and 10.8b have the same number of vertices. The vertices are connected identically except that graph G_2 in Fig. 11.8b has two fewer edges. We say that G_2 is a *subgraph* of G_1.

Finally, we come to the graph in Fig. 11.9. It has 6 vertices and 9 edges and the vertices in the top set of three are connected to each of the vertices in the bottom set of three, but no vertices of the top set are connected to one another, and no vertices of the bottom set are connected to one another. Such a graph is called a *complete*

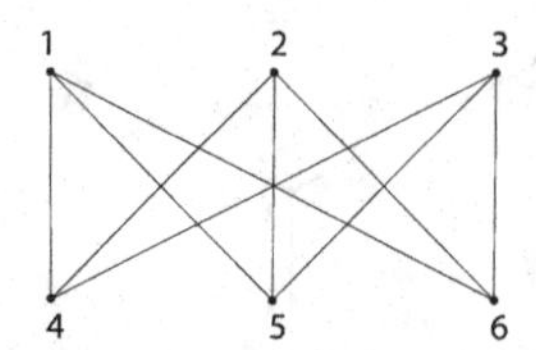

Fig. 11.9. Bipartite graph.

bipartite graph. Since there are 3 vertices in the top set and 3 vertices in the bottom set, the graph is symbolized by $K_{3,3}$. If there were 4 vertices in the top set and 3 in the bottom set, the bipartite graph would be symbolized by $K_{4,3}$. You can easily extend this to $K_{m,n}$. These graphs and their subgraphs will determine whether the framework is underbraced, minimally braced or overbraced. The next section shows how the rigidity of frameworks is related to bipartite graphs.

11.5. Determining Rigidity by Graph Theory

Consider 3×3 frameworks such as the ones shown in Fig. 11.3. Label the rows r_1, r_2 and r_3, and the columns c_1, c_2 and c_3 as shown in the framework in Fig. 11.10. Three of the frameworks in Fig. 11.3, namely the ones in Figs. 11.3d, 11.3h and 11.3a, correspond to graphs shown in Figs. 11.11a, 11.11b and 11.11c respectively. What we have done is to make r_1, r_2, r_3 and c_1, c_2, c_3 the vertices of the two sets of a subgraph of the bipartite graph $K_{3,3}$. Notice, in the framework of Fig.11.3d, there is a crossbar in cell (r_2, c_3). As a result, there is an edge in the corresponding bipartite graph between r_2 and c_3 as shown in Fig. 11.11a. In general, if there is an edge in the cell defined by row r_j and column c_k, there will be an edge between r_j and c_k in the corresponding subgraph of $K_{3,3}$. See that this is true by checking the locations of the crossbars in Figs. 11.3d, 11.3h and 11.3a corresponding to the edges of Figs. 11.11a, 11.11b and 11.11c respectively.

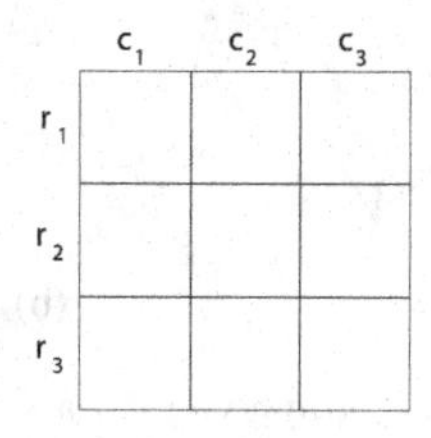

Fig. 11.10. A 3×3 grid defined in terms of rows and columns.

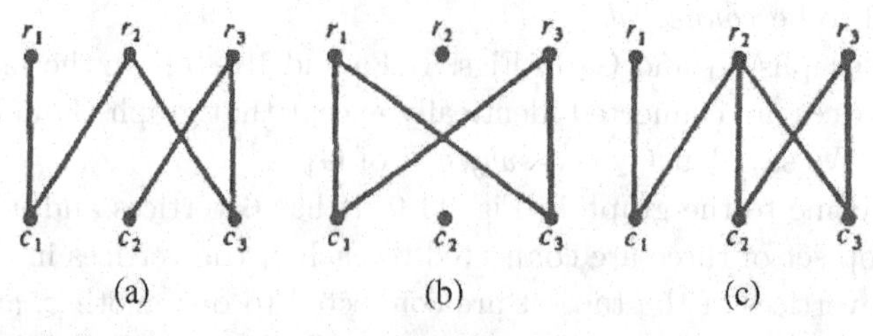

Fig. 11.11. Subgraphs of a bipartite graph illustrating a framework that is: (a) minimally braced; (b) underbraced; (c) overbraced.

Graph theory can now be used to determine if a framework is underbraced (not rigid), minimally braced or overbraced. The results are based on the following theorem a proof of which is sketched in Appendix 11A.

Theorem 11.1:

1. *A bracing of an* $n \times m$ *grid is rigid if and only if the corresponding bipartite subgraph is connected.*
2. *A bracing of an* $n \times m$ *grid is a minimal rigid bracing if and only if the bracing bipartite subgraph is a connected tree.*
3. *A bracing of an* $n \times m$ *grid is overbraced if and only if the bracing bipartite subgraph is connected and has a cycle.*

From this theorem, it follows that the framework in Fig. 11.3d is minimally braced (see Fig. 11.11a), the one in Fig. 11.3h is underbraced (see Fig. 11.11b), and the one in Fig. 11.3a is overbraced (see Fig. 11.11c). Also since a 3×3 framework has 6 vertices, the minimally braced framework, being a tree, will have 5 edges which is the necessary condition to minimally brace such a framework. However, merely having 5 braces does not guarantee rigidity. The braces must be properly placed.

Problems:

1. Use the graphical method to determine the rigidity of your frameworks from Sec. 11.2.
2. Apply the graphical method to evaluate the frameworks in Figs. 11.3a, 11.3*l*, and 11.3g.
3. Figure 11.12 is a 2×3 framework. Use the graphical method to show that this framework is underbraced. How many crossbars are needed to make the framework rigid? Which crossbars should be added?

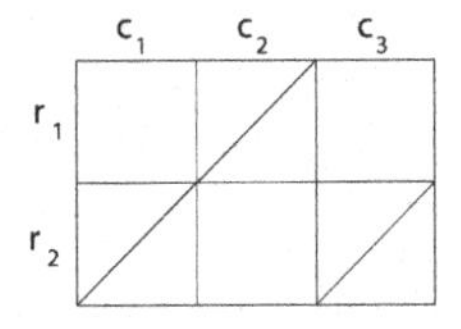

Fig. 11.12.

Appendix 11A. Proof of the Graphical Method for Determining the Rigidity of Frameworks

A sketch of the proof of Theorem 11.1 is given by the following an example. Consider the minimally rigid framework which is the result of the framework in Fig. 11.3d and its corresponding subgraph of the bipartite graph in Fig. 11.11a. Because Fig. 11.11a is a connected graph, there is a path from any row r to any column c. For example, there is a path from r_1 to c_3. I will demonstrate that the parallelogram in row r_1 and column c_3 must be a square, and by a similar argument, every element of the framework must be a square. So the framework must be rigid.

From 11.11a, we see that $r_1c_1r_2c_3$ is a path from r_1 to c_3. Referring to Fig. 11.3d and using the lemma, we see that the vertical edge of square (r_1, c_1) is perpendicular to the horizontal edge of (r_1,c_1), the horizontal edge of (r_1, c_1) is perpendicular to the vertical edge of (r_2, c_1), and the vertical edge of $(r_{2,}c_1)$ is perpendicular to the horizontal edge of (r_2, c_3). Therefore, the vertical edge of (r_1, c_1) is perpendicular to the horizontal edge of (r_2, c_3). By the lemma, the vertical edge of (r_1, c_3) must be perpendicular to the horizontal edge of (r_1, c_3). So (r_1, c_3) must be a square. A similar argument can be made to show that all parallelograms within a framework are squares so long as the graph is connected.

QED

CHAPTER 12

SIMILARITY

12.1. Similarity

Two figures are *similar* if they are congruent after a magnification or contraction as shown in Fig. 12.1. Figure 12.2 shows three measurements sharing the same proportion for a pair of similar stick figures:

$$\lambda = \frac{a'}{a} = \frac{b'}{b} = \frac{c'}{c} \tag{12.1}$$

where λ is the *magnification factor*.

12.2. Similar Triangles

The conditions for two triangles to be similar are:

Axiom 12.1: Two triangles are similar iff (if and only if) they agree in their three angles.

Axiom 12.2: Two triangles are similar if two of their sides are in the same proportion and their included angles are equal. This is called similarity by SAS.

If two triangles (see Fig. 12.3), ΔABC and $\Delta A'B'C'$ are similar, we denote this by: $\Delta ABC \sim \Delta A'B'C'$ and we are assured that:

$$\angle A = \angle A', \quad \angle B = \angle B', \quad \angle C = \angle C' \quad \text{and} \quad \frac{a'}{a} = \frac{b'}{b} = \frac{c'}{c}. \tag{12.2}$$

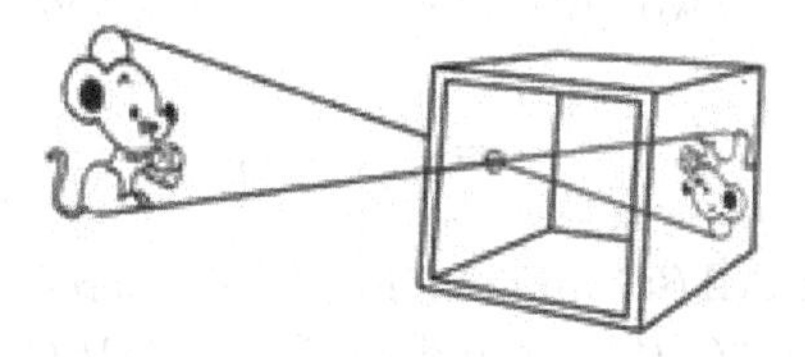

Fig. 12.1. A mouse projects to a scale model of himself.

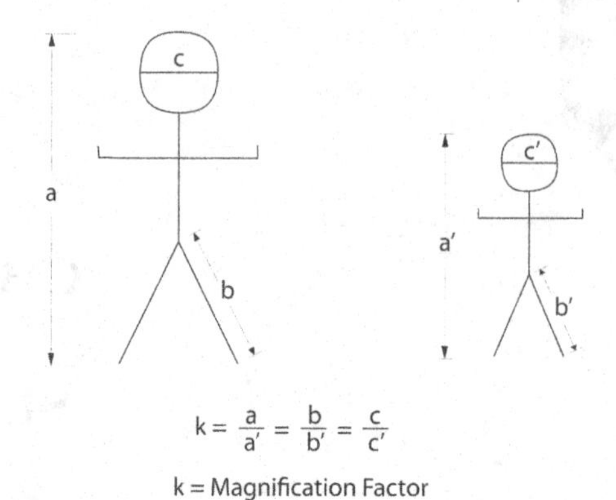

Fig. 12.2. Corresponding lengths of similar figures are in proportion.

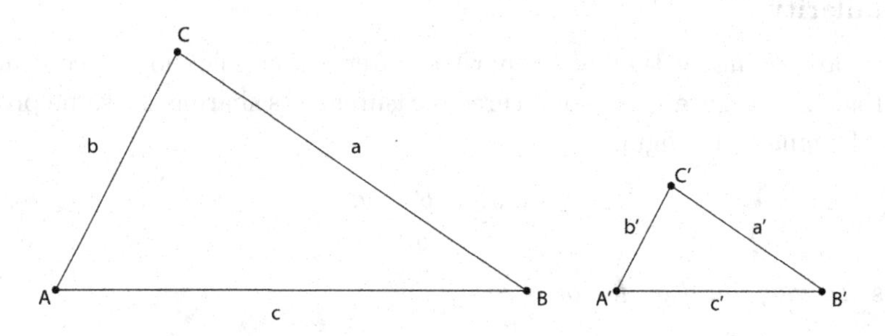

Fig. 12.3. Two similar triangles.

We have experienced Axiom 1 in Sec. 5.3.4 where, given three angles (AAA), numerous triangles can be constructed with these angles. However, each of these triangles is obtained from the others by magnification or contraction, and so they are all similar.

Axiom 2 gives us a justification for the Construction 4.2.3, Method 1 of a line through D parallel to the line segment AB (see Fig. 12.4). We used the following procedure:

a. Extend AD an equal distance to C.
b. Bisect line segment BC and denote by E, the point of bisection.
c. DE will be parallel to AB.

We now see why this works. Two of their sides are in proportion, i.e.,

$$\frac{AC}{DC} = \frac{BC}{EC} = 2, \tag{12.3}$$

and the included angle of ΔABC and ΔDEC is θ. Therefore, $\Delta ABC \sim \Delta DEC$ by SAS. As a result, $\angle CAB = \angle CDE = \alpha$ and $\angle ABC = \angle DEC = \beta$, and by the converse of Theorem 10.3, if complementary angles are equal, then AB and DE are parallel.

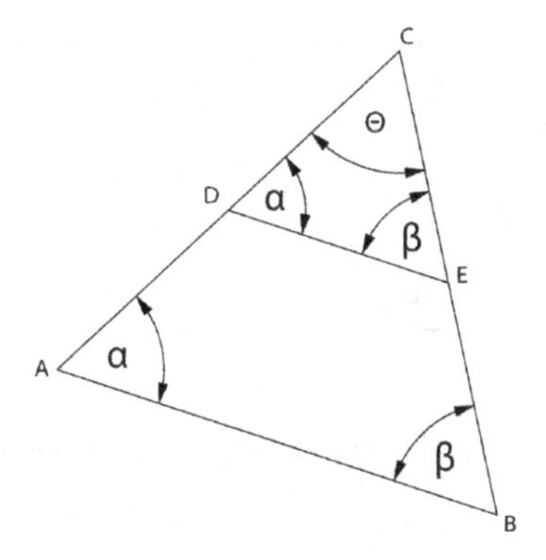

Fig. 12.4. Use of similar triangles to construct a line through D parallel to line segment AB.

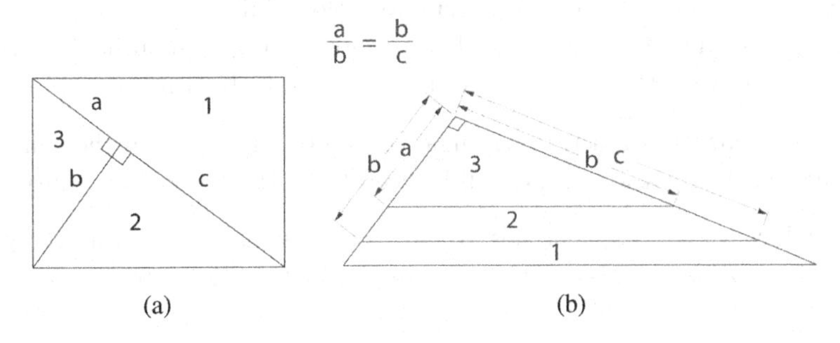

(a) (b)

Fig. 12.5. (a), (b) Surgery on a right triangle divides a rectangle into three similar triangles.

12.3. Surgery on a Right Triangle

Construction 1: Draw a diagonal on a rectangular piece of construction paper. Next, draw a line from a vertex of the rectangle, meeting the diagonal at right angles. Label the line segments meeting at the intersection of the diagonal, and the line from the vertex by the letters a, b, c as shown in Fig. 12.5a. Cut out the three triangles. This results in three similar triangles numbered $1, 2$ and 3 from large to small. Juxtapose the triangles so that their right angles are together and mark the sides of these triangles by their values a, b and c (see Fig. 12.5b). You can visually see that the triangles are similar and that,

$$\frac{a}{b} = \frac{b}{c}. \tag{12.4}$$

In other words, the small edge of Triangle 3 matches to the small edge of Triangle 2, as the large edge of Triangle 3 matches to the large edge of Triangle 2. In Fig. 12.6, the hypotenuses of Triangles $1, 2$ and 3 are labeled d, e and f, respectively.

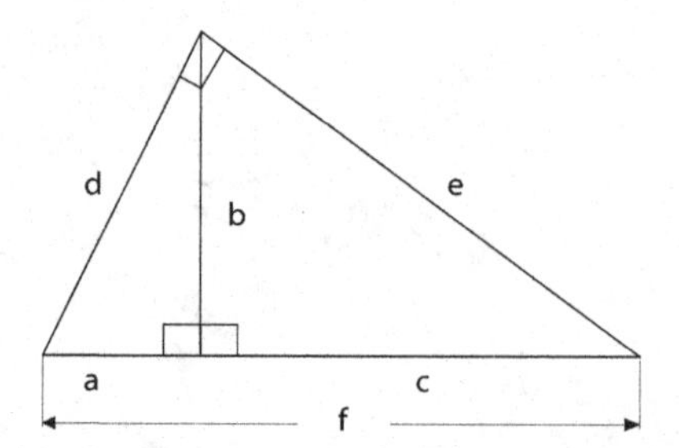

Fig. 12.6. A right triangle expressing the mean proportional of the altitude to the two legs on the hypotenuse.

Pythagoras was said to exclaim that the world has bequeathed us two great mathematical treasures:

a. The Pythagorean Theorem was a measure of gold, while
b. The construction in which the altitude of a right triangle, quantity b, is the mean proportional between the two legs a and c, was seen to be a measure of silver.

In Chap. 16, this surgery on a right triangle will be the first step in the generation of a logarithmic spiral. Equation (12.4) will be seen to be the key to this construction.

Problem 1: Equation (12.4) enables you to compute the sides of a triangle given certain partial information about it. For example, in Fig. 12.6, let $a = 4$ and $c = 9$. Find $b = x$.

Solution: Using Eq. (12.4),

$$\frac{4}{x} = \frac{x}{9} \quad \text{or} \quad x^2 = (9)(4) \quad \text{or} \quad x = 6.$$

Now that we know a, b and c in Fig. 12.6, the other edge lengths d, e and f can be determined by either similar triangles or the Pythagorean Theorem. Determine the lengths of the hypotenuses f, e and d of Triangles 1, 2, and 3, and show that they satisfy the Pythagorean Theorem for Triangle 1.

12.4. The Pythagorean Theorem

Consider again the three right triangles in Fig. 12.6. From the surgery on a right triangle in Sec. 12.3, we know that Triangles 1, 2 and 3 are similar. Therefore,

$$\frac{a}{d} = \frac{d}{f} \quad \text{and} \quad \frac{c}{e} = \frac{e}{f}. \quad \text{(Why?)}$$

Therefore,

$$d^2 = af \tag{12.5a}$$

and

$$e^2 = cf. \tag{12.5b}$$

Adding Eqs. (12.5a) and (12.5b),

$$d^2 + e^2 = f(a + c)$$

since $a + c = f$ as seen in Fig. 12.6,

$$d^2 + e^2 = f^2. \tag{12.6}$$

The Pythagorean Theorem has once again materialized, but this time in an algebraic context.

12.5. Chords, Secant Lines and Tangents to Circles

We proved in Theorem 8.4 that all inscribed angles of a circle that subtend the same arc are equal and equal to half of the central angle that subtends the same arc. From this theorem, along with the concept of similar triangles, we can prove the following theorem:

Theorem 12.1: *Given a circle and a pair of chords, AB and CD intersecting at point O within the circle,*

$$OA \times OB = OC \times OD. \tag{12.7}$$

Proof.

1. Referring to Fig. 12.7a, the inscribed angles α_1 and β_1 are equal since they intercept the same arc, BC of the circle.
2. By the same token, α_2 and β_2 are equal since they intercept arc AD.
3. $\alpha_3 = \beta_3$, since by Theorem 8.3, vertical angles are equal.
4. $\Delta COA \sim \Delta DOB$.
5. $\frac{OC}{OB} = \frac{OA}{OD}$.
6. $OA \times BO = OC \times OD$.

QED

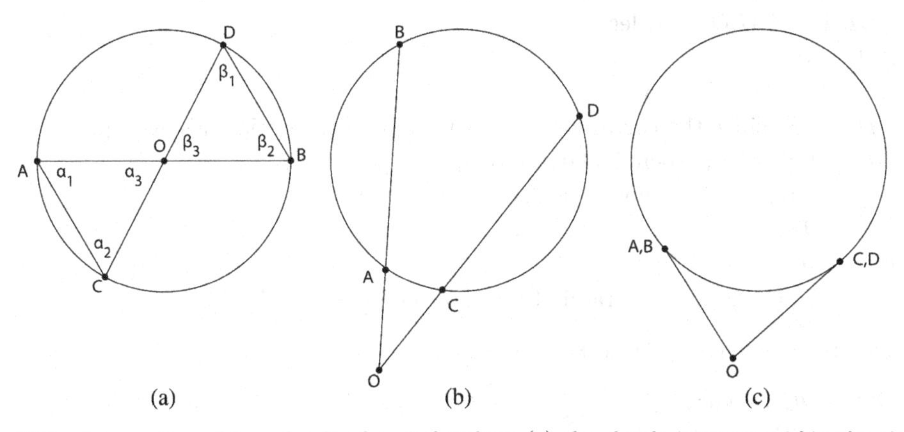

Fig. 12.7. Intersection of two chords of a circle where (a) the chords intersect within the circle; (b) outside the circle; and (c) the limiting case where the chords are tangent to the circle.

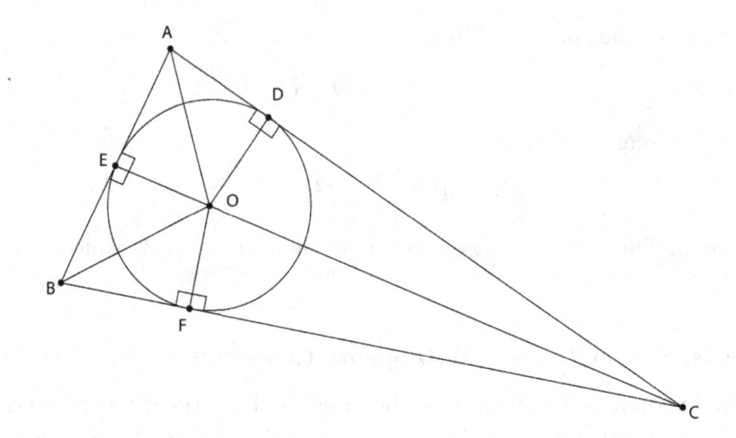

Fig. 12.8. Proof that the angle bisectors of a triangle meet at the center of the inscribed circle.

Two important corollaries follow from this theorem illustrated by Figs. 12.7b and 12.7c. The proofs are omitted.

Corollary 1: *If O is a point outside of the circle, and OAB and OCD are secant lines, $OA \times OB = OC \times OD$ still holds.*

Corollary 2: *If $A = B$ and $C = D$, and OA and OC are tangent to the circle, then $OA = OC$.*

Corollary 2 plays an important role in the next theorem.

Theorem 12.2: *The angle bisectors of ΔABC meet at point O which is the center of the inscribed circle* (*see Fig.* 12.8).

Proof.

1. Let O be the meeting point of the angle bisectors of $\angle CAB$ and $\angle ABC$.
2. $\angle CAO = \angle OAB$ and $\angle ABO = \angle OBC$.
3. From O, drop perpendiculars OD and OE to sides AC and AB of ΔABC.
4. $\angle ODA = \angle AEO = 90$ deg.
5. $AO = AO$.
6. $\Delta ODA \cong \Delta OAE$ by AAS.
7. $OD = OE$, since the corresponding parts of congruent triangles are equal.
8. From O, drop a perpendicular OF to BC.
9. Repeat Steps 1–7 to show that $\Delta OEB \cong \Delta OBF$.
10. $OF = OE$.
11. $OD = OE = OF$.
12. OD, OE and OF are the radii of the inscribed circle of ΔABC.

We need to show that OC bisects $\angle BCA$.

13. $OC = OC$ by identity.
14. CF and CD are tangent to the inscribed circle.
15. By Corollary 2 of Theorem 12.1, $CF = CD$.

16. $\Delta OFC \cong \Delta CDO$ by SSS (11, 13, 15).
17. $\angle FCO = \angle OCD$.

QED

12.6. Problems

2. Use the Pythagorean Theorem and the concept of similarity to solve for x in Figs. 12.9a–j. **Note:** The figures are not drawn to scale.
3. In Fig. 12.10 solve for a, b, c, d and e.
4. There is a famous puzzle attributed to Martin Gardner about a water lily that the poet Henry Longfellow introduced into his novel, "Kavenaugh." When the stem of

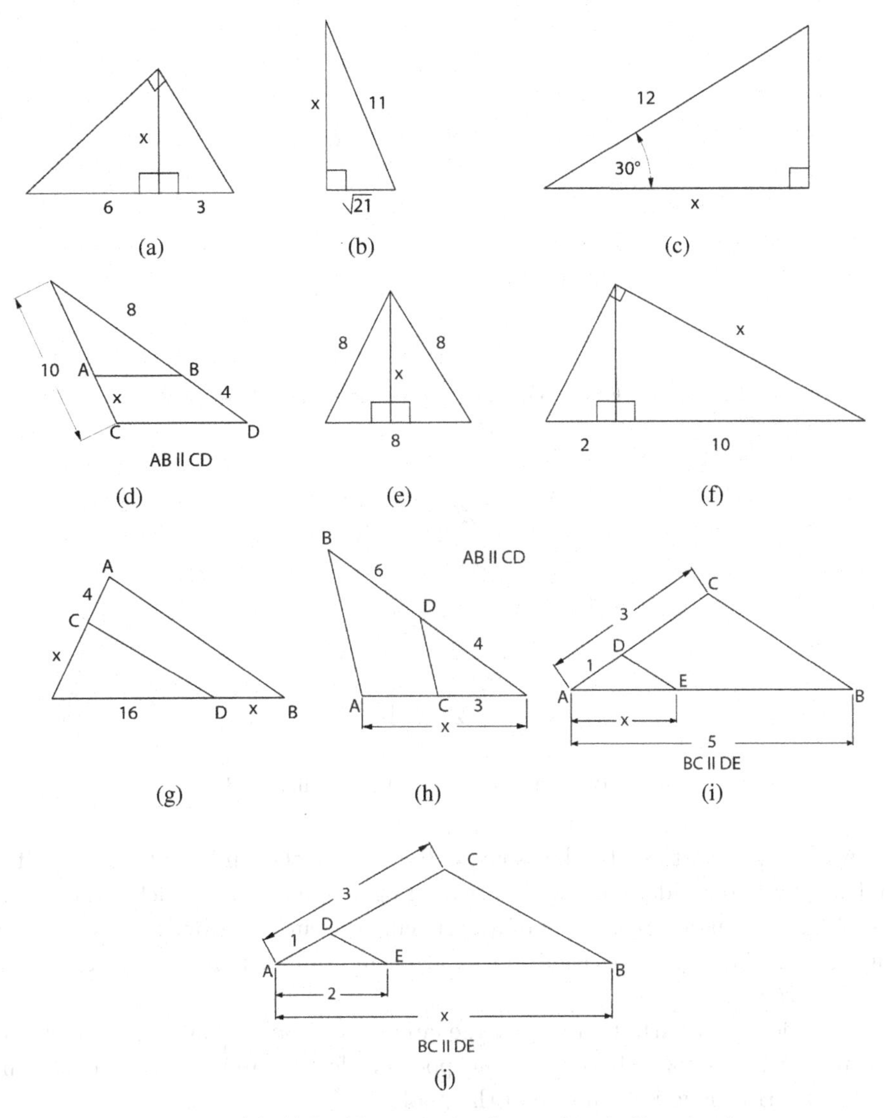

Fig. 12.9. (a)–(j) Problems to solve for x using similar triangles.

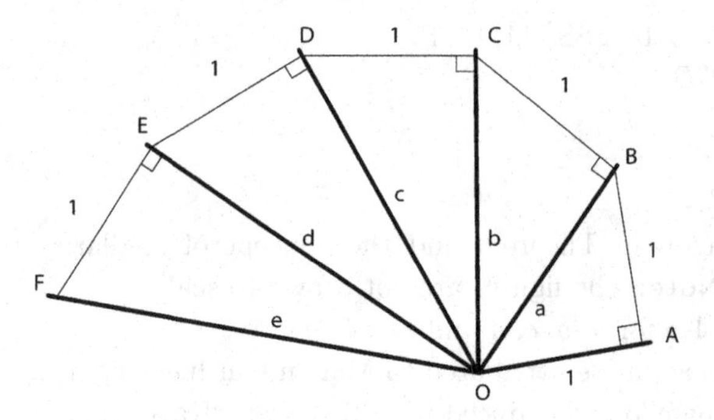

Fig. 12.10. Solve for a, b, c, d, e.

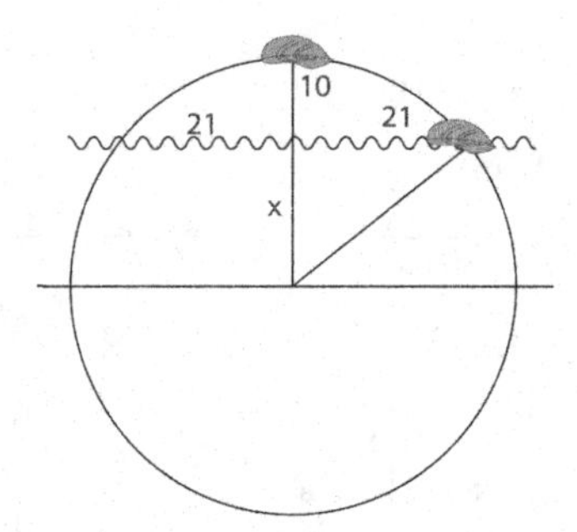

Fig. 12.11. From Longfellow's novel Kavenaugh. Solve for x.

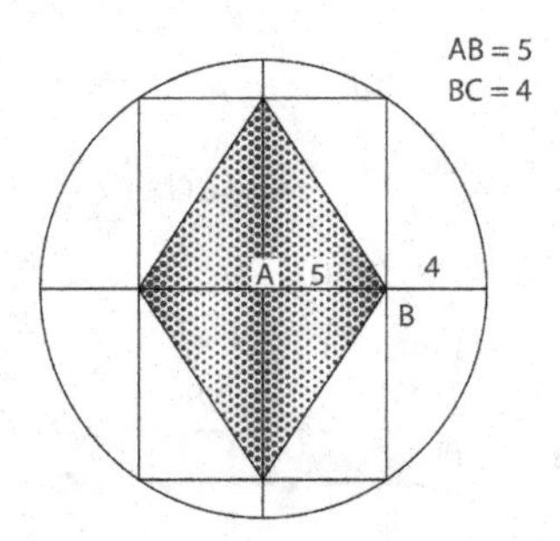

Fig. 12.12. Find the length of the wading pool.

the water lily is vertical, the blossom is 10 cm above the surface of the lake. If you pull the lily to one side, keeping the stem straight, the blossom touches the water at a spot 21 cm from where the stem formerly cuts the surface. How deep is the water? Figure 12.11 helps you to visualize this problem. Your task is to solve for x. Hint: Use Theorem 12.1.

5. In the middle of a park, there is a large circular play area. The city council would like to put a diamond-shaped wading pool inside the circular area, as shown in Fig. 12.12. How long is each side of the pool?
6. Prove the following theorem.

Theorem 12.3: *If the midpoints of the sides of a quadrilateral* (4*-sided polygon*) *are joined, they form a parallelogram. Hint*: *Draw the two diagonals of the quadrilateral and use similarity and Theorem* 10.3.

7. Suppose that one triangle is similar to another, and the ratio of corresponding sides is r, what can you say about the ratio of the areas? Prove your contention.
8. If an equilateral triangle has a side of length s, find the length of the altitude of the triangle in terms of s.
9. If a right triangle has an angle of 30 deg, find the ratio of the shorter side to the longer side.

CHAPTER 13

COMPASS AND STRAIGHTEDGE CONSTRUCTIONS

PART 2: DOING ALGEBRA WITH GEOMETRY

13.1. Introduction

This chapter will show, using the idea of similar triangles, how given two lengths 'a' and 'b' and a unit, '1', we can construct lengths:
$a+b$, $a-b$, $ka, ab, a/b$ and $\sqrt{a}$ using compass and straightedge, thereby reducing algebra and arithmetic to geometry.

13.2. Construction of $nl, a+b$, and $a-b$

Given a line segment l, it is obvious that segments of length $2l, 3l, 4l, \ldots$ can be created. Likewise, given two line segments a and b, it is equally clear how to construct segments $a+b$ and $a-b$.

Problem 1:

a. Choose your own length, l and construct $2l, 3l, 4l$ and $5l$.
b. Choose two lengths a and b, then construct lengths $a+b$ and $a-b$.

13.3. Construction of $\frac{a}{b}$ and ab

In Chap. 12, Problem 2i, we constructed $\frac{a}{b}$ where $a=5$ and $b=3$. The same construction of $\frac{a}{b}$ would work for any other values of a and b. In other words, we can construct $\frac{a}{b}$ by using the following algorithm (see Fig. 13.1):

Algorithm to construct $\frac{a}{b}$:

i. Mark off $OB=b$ and $OA=a$, incident to any angle at O.
ii. On side OB, mark off the unit length, $OD=1$.
iii. Through D, draw a line parallel to line AB intersecting line OA at C.
iv. The segment $OC=a/b$.

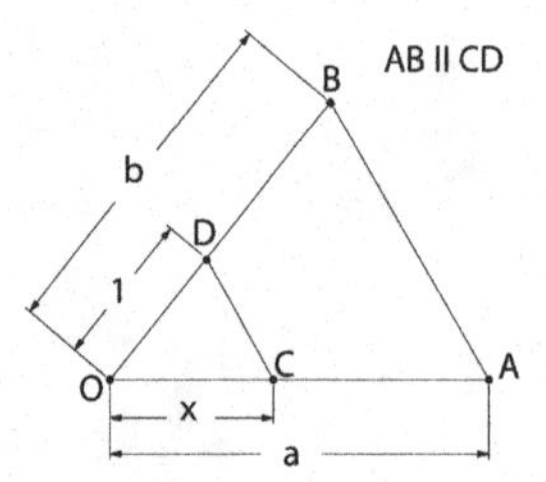

Fig. 13.1. Construction of the length $x = \frac{a}{b}$.

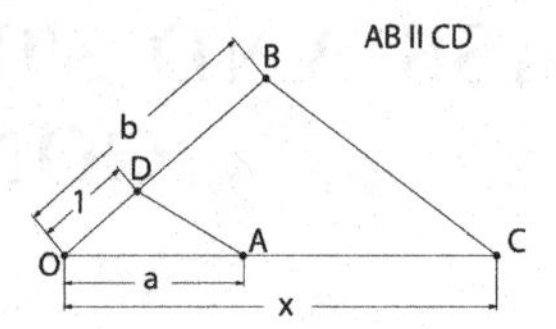

Fig. 13.2. Construction of $x = ab$.

Proof:

Let segment $OC = x$. Since ΔOCD and ΔOAB are similar (why?),

$$\frac{x}{a} = \frac{1}{b} \quad \text{or} \quad x = \frac{a}{b}.$$

Algorithm to construct ab:

In Problem 2j of Chap. 12, we constructed ab for $a = 2$ and $b = 3$.

The same construction of ab would work for any values of a and b using Fig. 13.2. Try to construct the algorithm and its proof on your own.

Problem 2:

Let 5 boxes on your graph paper be equivalent to 1 unit.

a. Using Fig. 13.1 with $a = 5$ units and $b = 3$ units, construct length a/b. Check your result by measurement.
b. Using Fig. 13.2 with $a = 3$ units and $b = 2$ units, construct length ab. Check your result by measurement.

13.4. Line Chopper

Given a length l, we have seen in Sec. 13.2 that you can create lengths of any integral multiple of l, i.e., l, $2l$, $3l$.... The length can also be divided into rational fractions of l by using the following line chopper.

a. A series of n equidistant line segments A_kB_k are drawn perpendicular to a given line drawn on graph paper as shown in Fig. 13.3 for $n = 3$.
b. For line segment l, take your compass and transfer l so that one end is at point A_0 and the other end intersects line A_nB_n of the chopper. Draw this line segment which has the same length as l.

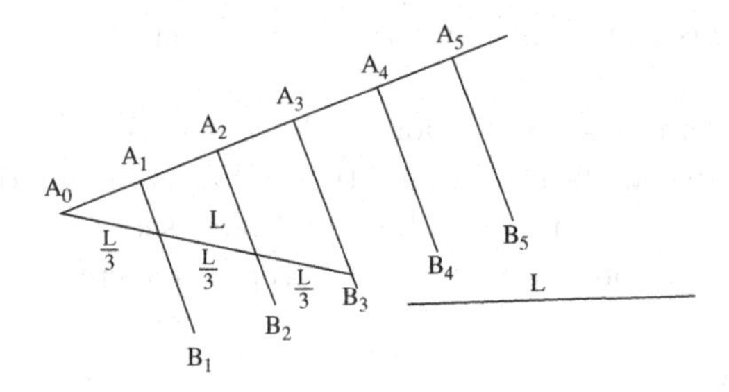

Fig. 13.3. A line chopper illustrated for $\frac{1}{3}l$ and $\frac{2}{3}l$.

c. Where this line segment intersects the other lines from the chopper, l is divided into n equal parts. For example, in Fig. 13.3, as a result of similar triangles, l is divided into three equal parts. In this way, fractional units of lengths $\frac{1}{3}l$, $\frac{2}{3}l$ can be constructed.

d. Proceeding in this manner, any rational length $\frac{p}{q}l$ can be constructed where p and q are integers.

Remark: It is important that l be longer than the line of the line chopper from A_0 up to A_n. If l is *not* longer, then you must decrease the spacing of lines A_kB_k.

Problem 3:

Choose a line segment of length l and use a line chopper to create lengths: $\frac{3}{5}l$, $\frac{2}{7}l$, $\frac{5}{8}l$, i.e., $n = 5, 7$ and 8 on the line chopper and mark them off on length l. Make sure that l is longer than the distance from A_0 to A_n.

13.5. Doing Algebra with Geometry

Beginning with a unit and lengths a and b, in Sec. 13.2, we constructed $a \pm b$; in Sec. 12.3 we constructed $\frac{a}{b}$ and ab. Now let us construct $\frac{a+b}{a-b}$ and $(a+b)(a-b) = a^2 - b^2$.

To do this we use the algorithm to construct $\frac{a}{b}$, but where we previously placed 'a', we now place $a + b$, and where we previously placed 'b', we now place $a - b$. The same goes for the construction of $(a+b)(a-b) = a^2 - b^2$.

Problem 4:

With a unit equal to 6 graph paper boxes, construct lengths $a = 2\frac{3}{7}$ units and $b = \frac{3}{5}$ unit and use them to construct the following lengths:

a. $a + b$, $a - b$

b. $\frac{a+b}{a-b}$

c. $(a+b)(a-b) = a^2 - b^2$.

To construct 'a', measure out a horizontal line segment with 2 units (12 boxes). Then use your line chopper to create an additional length of $\frac{3}{7}$ unit. Similarly, $\frac{3}{5}$ unit is gotten by cutting the length of the unit (length of 6 boxes) down to $\frac{3}{5}$ unit by using the line chopper.

Use arithmetic to compute these quantities and check the result of your construction by measurement.

From these considerations, any "rational" algebraic expression can, in principle, be constructed by successive application of addition, subtraction, multiplication and division. More complex constructions can be built up from a set of given lengths $a, b, c, \ldots$. For example, we can now construct $a^2 - b^2$, $\frac{a+b}{a-b}$, $3a^2 - ab$, etc.

13.6. Square Roots

A decisive new construction that carries us beyond the "rational" field is the construction of a square root.

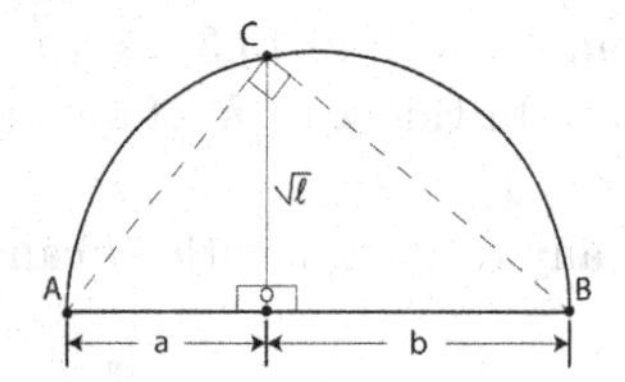

Fig. 13.4. Construction of $\sqrt{l}$.

a. To construct $\sqrt{l}$, we refer to Fig. 13.4.
   i. Let l be the product of two numbers, i.e., $l = ab$.
   ii. On a straight line, we mark off $OA = a$ units and $OB = b$ units.
   iii. Draw a circle with AB as the diameter. (You will have to use Construction 1 of Sec. 4.2.1 to bisect segment AB in order to find the radius of the circle).
   iv. Construct the perpendicular line to diameter AB through O until it meets the circle at C.
   v. By the corollary to Theorem 8.4, $\angle ACB$ is a right angle; hence the right triangles ΔAOC and ΔCOB are similar. Thus by Eq. (12.4), where $AC = x$,

$$\frac{a}{x} = \frac{x}{b} \quad \text{or} \quad x^2 = ab = l \quad \text{or} \quad x = \sqrt{l}.$$

b. If you wish to construct the square root of an integer, there is an easier way to do this. It is illustrated in Fig. 13.5 and uses the Pythagorean Theorem. This construction was introduced in Problem 3 of Chap. 12.

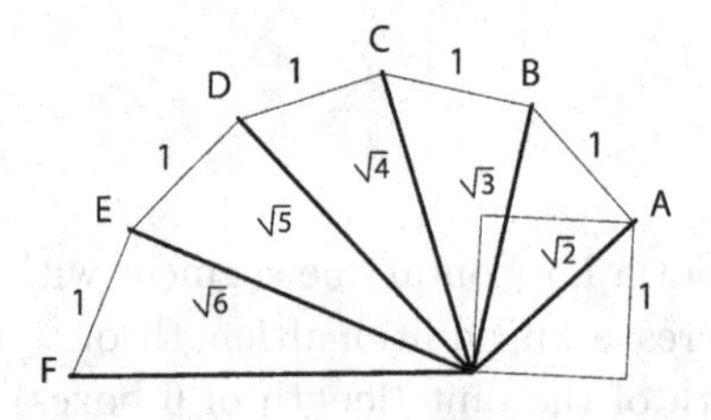

Fig. 13.5. Consecutive constructions of $\sqrt{n}$.

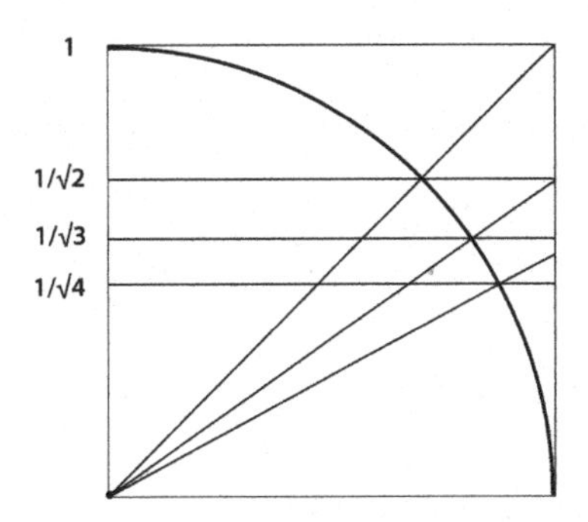

Fig. 13.6. Construction of consecutive rectangles with proportion 1: $\sqrt{n}$ for positive integers n.

We will use $\sqrt{l}$ in Chap. 16 to construct a dense set of points on a logarithmic spiral.

Problem 5: (a) With a unit equal to 4 boxes, construct $\sqrt{2.5}$ units.
(b) With a unit of 2 boxes construct $\sqrt{6} = \sqrt{(3)(2)}$.

13.7. Construction of a Sequence of Rectangles with Proportions $\sqrt{n} : 1$

The following is an ancient method for creating a sequence of rectangles with proportion 1: $\sqrt{n}$ (see Fig. 13.6).

a. Begin with a square 1:1.
b. Place the right-leaning diagonal in the square.
c. From the lower left-hand vertex of the square, sweep the arc of a quarter circle with radius equal to the length of the side of the square.
d. Where the arc intersects the diagonal, draw a horizontal line until it intersects the right side of the square forming a rectangle. This rectangle has proportions, $\sqrt{2} : 1$.
e. Draw the diagonal of this rectangle. Where it intersects the arc of the circle, draw a horizontal line forming another rectangle. This rectangle has proportions $\sqrt{3} : 1$.
f. Continue in this manner to create rectangles of proportion $\sqrt{n} : 1$ for $n = 1, 2, 3, \ldots$.

[illegible] sequence of [illegible] with proportion [illegible]

[illegible]

[illegible] Construction of a Sequence of Rectangles with Proportions [illegible]

[illegible] the following construction for generating a sequence of rectangles with proportions [illegible]

[illegible]
[illegible] the diagonal [illegible] the square.
[illegible] from the lower left and [illegible] of the square [illegible] the arc of a circle [illegible] equal to the length of the side of the square.
[illegible] the arc intersects the diagonal, draw a horizontal line [illegible] right side of the square to form a rectangle. [illegible]
[illegible] diagonal [illegible] this rectangle [illegible]
[illegible] horizontal line [illegible]
[illegible] Continue in this manner [illegible] rectangles [illegible] proportions [illegible]

CHAPTER 14

AREA

14.1. Introduction

The concept of area follows from four common-sense axioms one of which tells us how to find the area of a rectangle. Everything that we can say about areas follows by deduction. This is an excellent example of how the deductive method can be used to derive complexity from simplicity. We will derive an important property of the area of families of similar figures and use it to present a wonderfully simple proof of the Pythagorean Theorem. We will explore area with the use of a tool that used to be revered greatly in the lower grades called a geoboard. In this and the next chapter we will see how geoboards are a perfect gameboard to express advanced concepts of geometry within a playful context. First, we state the four axioms of plane areas.

14.2. Axioms of Area

Axiom 14.1: To every polygonal region there corresponds a definite positive real number called its area.

Axiom 14.2: If two triangles are congruent, they have the same area.

Axiom 14.3: If a polygonal region R is the union of two polygonal regions R_1 and R_2 with the boundaries shared between them at most a finite number of line segments, then the area of R is the sum of the areas of R_1 and R_2, i.e.,

$$\text{Area }(R) = \text{Area }(R_1 \cup R_2) = \text{area }(R_1) + \text{area }(R_2)$$

Axiom 14.4: The area of a rectangle (a four-sided figure whose angles all measure 90 deg.) is the product of the length of its base and its height.

The area of every polygonal shape can be derived from these axioms. Since any region in the plane can be approximated by many-sided polygons, these axioms can be extended beyond polygons to more general regions by using the elements of calculus.

Let us now apply these axioms to find the area of some simple geometric figures.

14.3. Area of a Parallelogram

Subdivide parallelogram $ABCD$ as shown in Fig. 14.1a (Axiom 14.3). Transform the parallelogram in Fig. 14.1a to Fig. 14.1b. We now have two rectangles, one with base b_1 and one with base b_2, both having height h.

As a result,

$$A = b_1 h + b_2 h = (b_1 + b_2)h$$

where,

$$b_1 + b_2 = b$$

or

$$A_{\text{par}} = bh \tag{14.1}$$

by Axiom 14.4. Therefore, the area of a parallelogram is the length of its base times its height.

14.4. Area of a Triangle

Since two triangles juxtaposed make a parallelogram (see Fig. 14.2), the area of a triangle is given by the formula,

$$A_{\text{tri}} = \frac{1}{2}bh. \tag{14.2}$$

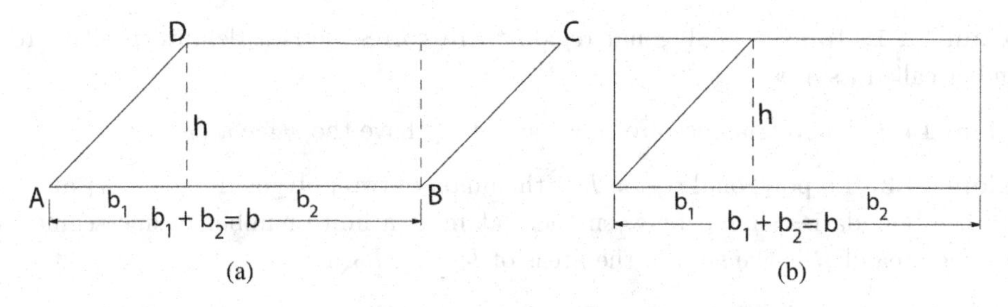

Fig. 14.1. (a) Parallelogram; (b) parallelogram transformed to a rectangle.

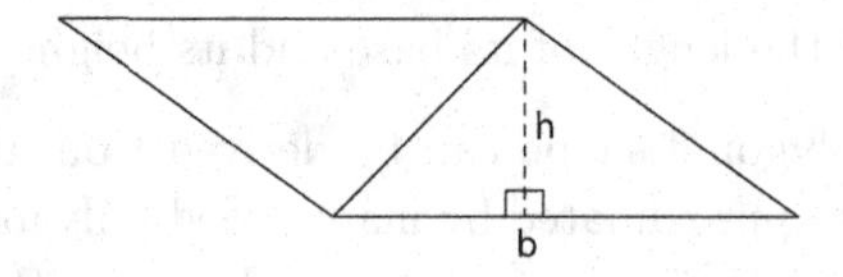

Fig. 14.2. Parallelogram transformed to a triangle.

If you know the lengths a, b, c of the sides of a triangle, i.e., SSS, you can use a formula attributed to Heron of Alexandria (10–70 A.D.), a Greek engineer and mathematician:

$$A = \sqrt{s(s-a)(s-b)(s-c)} \tag{14.3}$$

where s is the semi-perimeter. For example, a 3,4,5-right triangle has perimeter equal to 12 units so that using Eq. (14.3),

$$A = \sqrt{6(6\text{–}3)(6\text{–}4)(6\text{–}5)} = 6.$$

14.5. Area of a Polygon

Since any polygon can be subdivided into triangles, the area of a polygon is the sum of the areas of its subdivided triangles as shown in Fig. 14.3. Since any region bounded by a smooth curve can be approximated by a polygon, once you are able to find the area of a triangle, you can approximate the area within any smooth closed curve as accurately as you wish. In this way, Archimedes was able to approximate the area of a circle of radius 1, which equals π by considering inscribed and circumscribed 96-gons.

14.6. Area of a Trapezoid

A trapezoid is a four-sided figure with two parallel edges of unequal length as shown in Fig. 14.4. The parallel edges have length b_1 and b_2 while the height is h. Again, we subdivide the trapezoid according to Axiom 3 as shown in Fig 14.4.

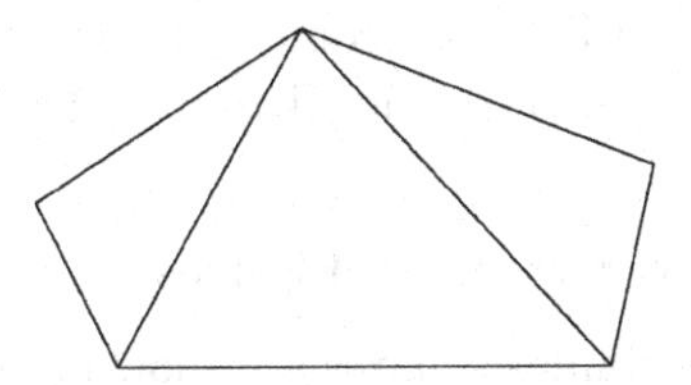

Fig. 14.3. Polygon subdivided into triangles.

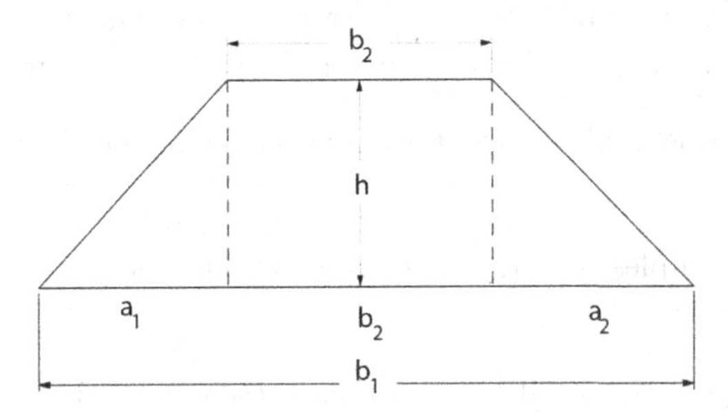

Fig. 14.4. Trapezoid subdivided into a rectangle and two right triangles.

Notice that,

$$b_1 = a_1 + b_2 + a_2$$

and,

$$
\begin{aligned}
A_{\text{trap}} &= \frac{h}{2}a_1 + hb_2 + \frac{h}{2}a_2 \\
&= \frac{h}{2}(a_1 + 2b_2 + a_2) \\
&= \frac{h}{2}((a_1 + b_2 + a_2) + b_2) \\
A_{\text{trap}} &= \frac{h}{2}(b_1 + b_2). \qquad (14.4)
\end{aligned}
$$

So we have derived the area of a trapezoid as the product of its height and the average of its bases from the axioms.

14.7. Area of an Equilateral Triangle

An altitude to the base of an equilateral triangle divides the triangle into two 30,60,90-triangles as shown in Fig. 4.5a. It is important to know that the edge lengths of a 30,60,90-triangle are given in Fig. 4.5b. As a result, if the edge of the equilateral triangle is taken to be 1 unit, its area is,

$$A = \frac{\sqrt{3}}{4}. \qquad (14.5)$$

Make sure that you understand how this computation was made. We used a triangle similar to the one in Fig. 4.7b in which the edges were half as large.

14.8. Areas of a Family of Similarity Figures

Three members of a family of similar squares are shown in Fig. 14.5a with side lengths in progression: $l_1 = l, l_2 = 2l, l_3 = 3l, \ldots$. It is clear from Fig. 14.5a that the areas increase progressively: $A_1 = A, A_2 = 4A, A_3 = 9A, \ldots$ (show this). In Fig. 14.5b, the first three members of a family of similar triangles are shown with side lengths in progression: $l, 2l, 3l, \ldots$. Again the areas are shown to be in progression: $A, 4A, 9A$.

Remark: Although this is illustrated for equilateral triangles, it holds for any family of similar triangles.

Both of these examples illustrate a general principle for the area of similar plane shapes stated as follows:

Governing Principle: For similar figures, as a characteristic length l is multiplied k-fold, its area multiplies k^2-fold.

This is the governing principle for all families of similar geometric figures. So in Fig. 14.5c are shown three similar sombreros from a family of sombreros. A characteristic length is chosen and the first two members of a family of similar sombreros

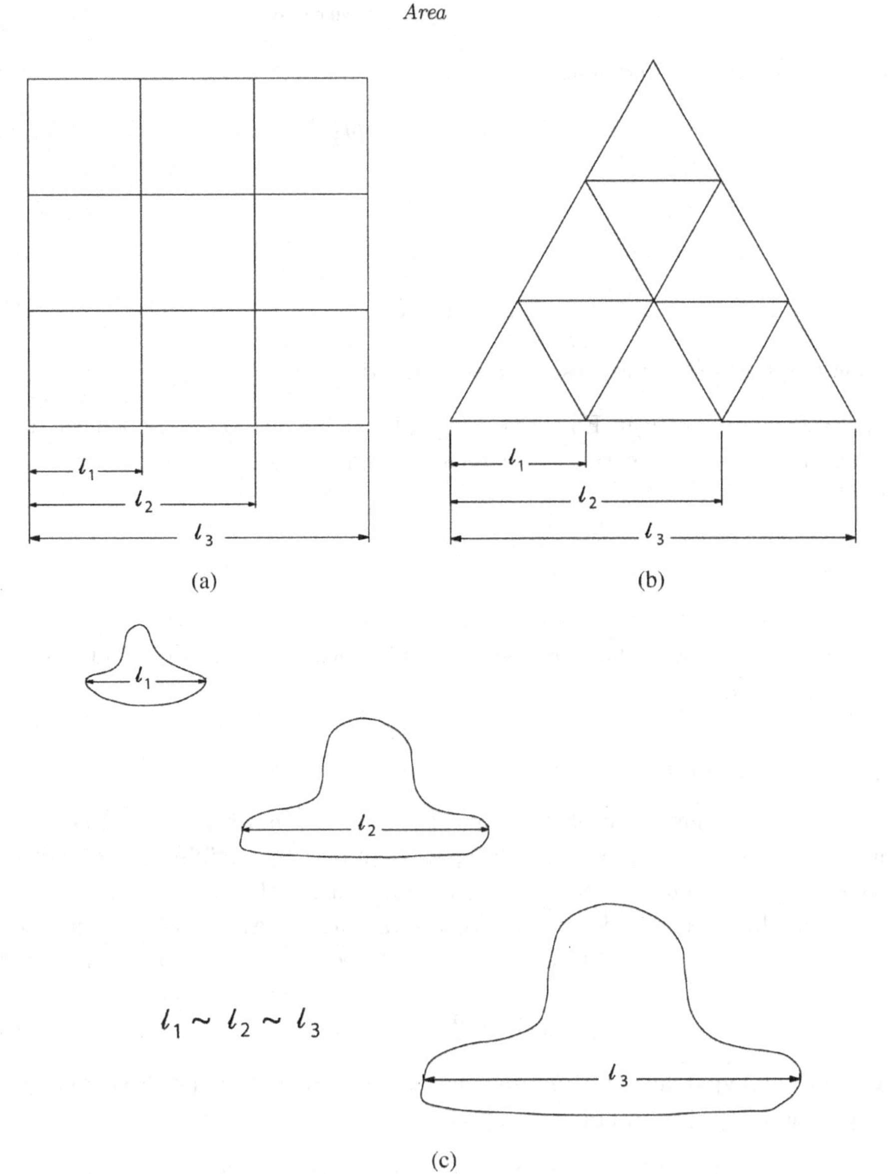

Fig. 14.5. (a) A progression edge lengths: $l_1 = 1, l_2 = 2, l_3 = 3$ for similar: (a) squares; (b) triangles; (c) sombreros.

are shown with edge lengths in progession $l, 2l, 3l$. Again the areas are in progression: $A, 4A, 9A$. We can restate this governing principle as follows:

For similar figures, the area is directly proportional to the square of any characteristic length, l, which is symbolized by,

$$A \propto l^2.$$

This expression can be rewritten as an equation:

$$A = kl^2. \tag{14.6}$$

Using this equation, we find that,

$$A_1 = kl_1^2, \quad A_2 = kl_2^2 \tag{14.7a,b}$$

Dividing Eq. (14.7b) by Eq. (14.7a),

$$\frac{A_2}{A_1} = \left(\frac{l_2}{l_1}\right)^2. \tag{14.8}$$

Equation (14.8) clearly expresses the governing principle.

Remark: The value of k in Eq. (14.6) depends on the characteristic length chosen. For example, if we choose the characteristic length of a square to be the side s, Eq. (14.6) becomes,

$$A = s^2.$$

Therefore $k = 1$.

If the characteristic length is the diagonal, since $s = \frac{d}{\sqrt{2}}$, Eq. (14.6) becomes $A = \frac{1}{2}d^2$, therefore $k = 1/2$.

14.9. Pythagorean Theorem

I will now state what is probably the simplest and most elegant proof of the Pythagorean Theorem. A right triangle T_1 with an altitude dividing it into two right triangles T_2 and T_3 similar to the original is shown in Fig. 14.6.

That the three triangles form a family of similar triangles follows from the surgery on a right triangle carried out in Chap. 12. The three areas clearly satisfy Eq. (14.9),

$$A_1 = A_2 + A_3. \tag{14.9}$$

Choose the hypotenuse h to be the characteristic length of the three triangles so that replacing Eq. (14.6) in (14.9),

$$kh_1^2 = kh_2^2 + kh_3^2.$$

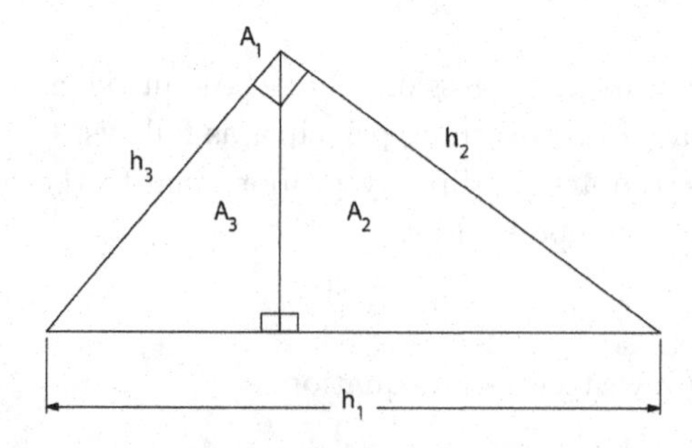

Fig. 14.6. The Pythagorean theorem follows from a family of three similar right triangles.

Dividing through by k,

$$h_1^2 = h_2^2 + h_3^2,$$

results in the Pythagorean Theorem.

14.10. Geoboards

14.10.1 Areas of polygons on geoboards by counting squares

A geoboard is an $m \times m$ grid of pins. For example, a 5×5 geoboard is shown in Fig. 14.7. Using the geoboard, areas can be found by counting squares.

A single square, shown in Fig. 14.7, has an area of 1 square unit.

For the rectangle shown in Fig. 14.8a, you can find its area by simply counting squares, i.e., $A = 8$ squares. To find the area of the triangle in Fig. 14.8b, embed it in a rectangle. Find the area of the rectangle (8 squares) and divide it in half; i.e., $A = \frac{1}{2}(8) = 4$ square units.

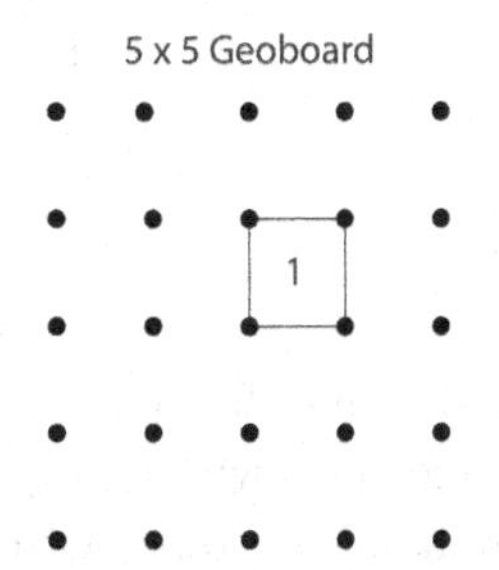

Fig. 14.7. A 5×5 geoboard with one unit of area highlighted.

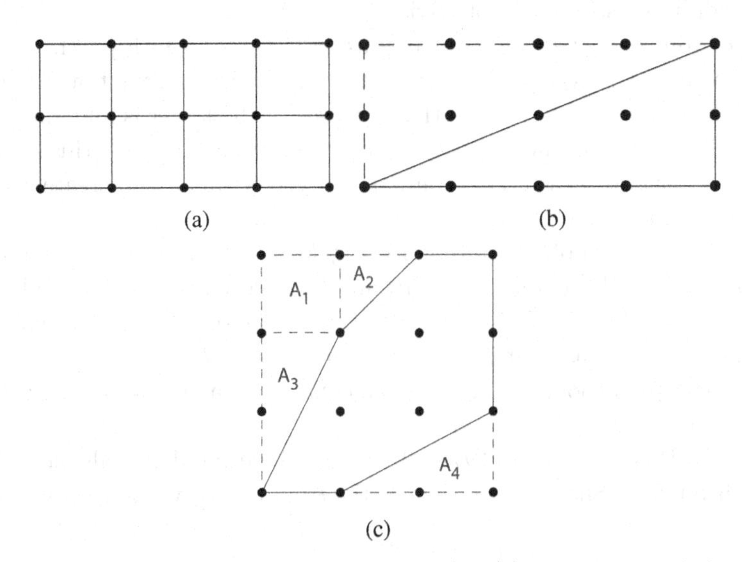

Fig. 14.8. (a) Area of a rectangle found by counting squares; (b) Area of a triangle found by embedding it in a rectangle; (c) Area of a polygon found by embedding it in a rectangle.

In general, you can apply the method of counting squares to an arbitrary polygon by first enveloping the polygon in a rectangle. For example, to find the area of the polygon in Fig. 14.8c, envelop the polygon in a rectangle as shown. The area of the polygon is found by subtracting from the area of the rectangle the areas of the triangles and rectangles within the rectangle and outside of the polygon. In this example,

$$A = \text{A rect} - (A_1 + A_2 + A_3 - A_4) = \left(9 - 1 - \frac{1}{2} - 1 - 1\right) = \frac{11}{2}$$

14.10.2 Areas on geoboards by the use of Pick's Law

For each region you can find the area by using Pick's Law which states

$$A = I + \frac{C}{2} - 1 \tag{14.10}$$

where I is the number of lattice points (upright pins on the geoboard) entirely within the boundary of the polygon, and C is the number of lattice points on the circumference or perimeter. For example, in Fig. 14.8c,

$$I = 3,\ C = 7, \quad \text{and} \quad A = 3 + \frac{7}{2} - 1 = \frac{11}{2}$$

which checks with the previous result.

14.11. Problems

1. How many squares of different side length can you find on a 5×5 geoboard? Sketch each of your squares on a 5×5 grid drawn on graph paper.
2. Take a geoboard and try your hand at constructing three polygons and compute their areas by counting squares as we did in Sec. 14.10.1 above. Be bold and make your polygon interesting and not trivial.
3. For each of your areas verify the area by using Pick's Law in Eq. (14.10).
4. Create two similar polygons. The larger one should have edges that are twice the length of the smaller. Check to see that the area of the larger is four times the area of the smaller in accordance with the "governing principle." For this exercise you can use a larger grid drawn on graph paper if you wish. Again, create a polygonal shape that is not trivial.
5. Is it true that the area of a quadrilateral is $\frac{1}{2}(b_1h_1 + b_2h_2)$ where b_1 and b_2 are the lengths of any two adjacent sides, and h_1 and h_2 are the lengths of altitudes to those sides from the vertex that does not lie on either of those sides? If it is not true, give an example to show this. If it is true, prove it.
6. Show that if a parallelogram has all sides equal (i.e., a rhombus), then its area is one-half the product of its diagonals.
7. Suppose $ABCD$ is a square and E is the point where the diagonals meet. Let L be any line through E. Show that L divides the square into two equal areas.

CHAPTER 15

VECTORS AND GEOBOARDS

15.1. Introduction

The concept of a vector is closely associated with geometry. Many theorems of geometry can be proven by the use of vectors. Vectors can also simplify trigonometric computations, offer another way to compute areas, provide an essential ingredient to an understanding of symmetry, and establish connections between geometry and physics.

15.2. What is a Vector?

A *vector* $\vec{v}$ is a quantity with *magnitude* and *direction.* It can be visualized by an arrow displayed on a geoboard. For example in Fig. 15.1, five vectors are shown: $\vec{v}_1, \vec{v}_2, \vec{v}_3, \vec{v}_4, \vec{v}_5$.

Each vector has a tail and a tip. The tip is distinguished by having an arrowhead attached to it while the tail has no arrowhead. A vector is named by taking a walk from tail to tip along the geoboard grid, going left or right and up or down. For example, for $\vec{v}_1$ from tail to tip: go 2 units to the right and 1 unit up. This can be represented by the number pair or *coordinates*, (2, 1). For vector $\vec{v}_2$: go to the left 1 unit and up 3 units, i.e., (–1, 3). Notice that vector $\vec{v}_3$ is clearly represented by (2, 1) so that we see that $\vec{v}_3$ is identical to $\vec{v}_1$ even though the tail of $\vec{v}_3$ is at a different location than the tail of $\vec{v}_1$. Although a vector *depends on its direction and magnitude, it is independent of the location of its starting point (tail).* Therefore, vectors with the same name are parallel to each other.

Next consider $\vec{v}_4$. From tail to tip: go 2 units to the left and 1 unit down or $\vec{v}_4 = (-2, -1)$. Notice that $\vec{v}_1$ and $\vec{v}_4$ are identical except that their tails and tips are interchanged. We also say $\vec{v}_4 = -\vec{v}_1$.

Remark: A negative value of the components of a vector represents a move to the left or down whereas positive values refer to moves to the right or up.

15.3. The Magnitude of a Vector

We have seen in Sec. 15.2 how to represent the direction of a vector. The magnitude of a vector can be computed by the Pythagorean Theorem. The *magnitude* of a vector

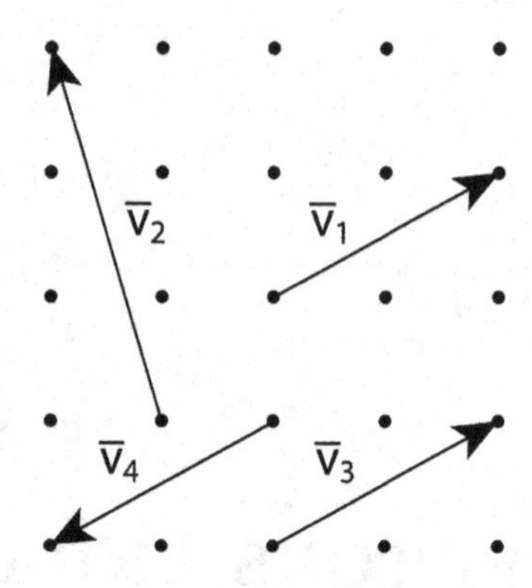

Fig. 15.1. Four vectors on a geoboard.

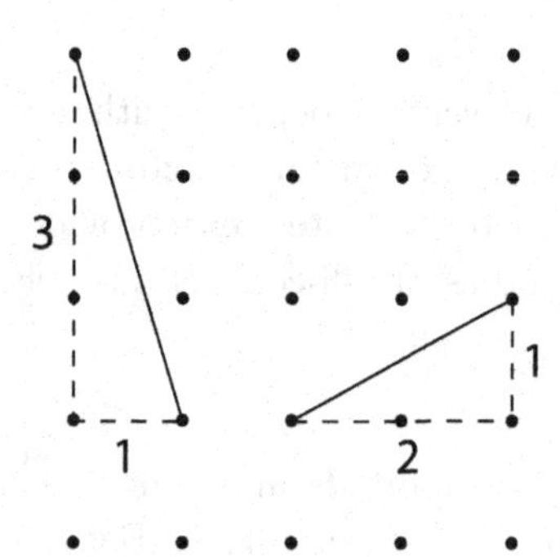

Fig. 15.2. Finding the length of two vectors.

is its length. You can find the length of a vector by using the Pythagorean Theorem. For example, the vector (2, 1) defines a right triangle, shown in Fig. 15.2 with dotted lines. The base has length 2 and height 1 so, by the Pythagorean Theorem, the length of (2, 1), designated by $|\vec{v}_1| = |(2,1)|$ is (see Fig. 15.2),

$$|\vec{v}_1| = |(2,1)| = \sqrt{2^2 + 1^2} = \sqrt{5}. \tag{15.1}$$

Also in Fig. 15.2,

$$|\vec{v}_2| = |(-1,3)| = \sqrt{(-1)^2 + 3^2} = \sqrt{10}. \tag{15.2}$$

In general the length of vector $\vec{a} = (a_1, a_2)$ is,

$$|(a_1, a_2)| = \sqrt{a_1^2 + a_2^2}. \tag{15.3}$$

15.4. The Sum and Difference of Vectors

In this section two operations will be defined on the pair of vectors $\vec{v}_1$ and $\vec{v}_2$ from Fig. 15.1.

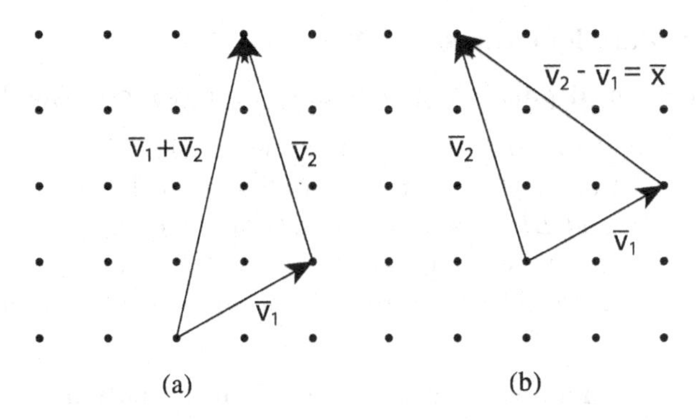

Fig. 15.3. Finding for two vectors, (a) their sum; (b) their difference.

15.4.1 The sum of two vectors $\vec{v}_1 + \vec{v}_2$

Take $\vec{v}_2$ and move it so that the tail of $\vec{v}_2$ touches the tip of $\vec{v}_1$. The *sum* will be the vector from the tail of the $\vec{v}_1$ to the tip of $\vec{v}_2$. For example, in Fig. 15.3a we see that $\vec{v}_1 + \vec{v}_2 = (1, 4)$. We also see that it is easy to compute the sum of two vectors by adding their corresponding coordinates, i.e., $\vec{v}_1 + \vec{v}_2 = (2, 1) + (-1, 3) = (1, 4)$.

In general, if $\vec{u} = (a_1, a_2)$ and $\vec{v} = (b_1, b_2)$, then

$$\vec{u} + \vec{v} = (a_1, a_2) + (b_1, b_2) = (a_1 + b_1, a_2 + b_2). \tag{15.4}$$

15.4.2 The difference between two vectors $\vec{v}_2 - \vec{v}_1$

Juxtapose vectors $\vec{v}_1$ and $\vec{v}_2$ from Fig. 15.1 so that their tails are together as in Fig. 15.3b, and find the vector from the tip of $\vec{v}_1$ to the tip of $\vec{v}_2$. This vector is $\vec{v}_2 - \vec{v}_1$. Notice that $\vec{v}_2 - \vec{v}_1 = (-3, 2)$. Again, we can compute the *difference* of two vectors by subtracting their coordinates, i.e.,

$$\vec{v}_2 - \vec{v}_1 = (-1, 3) - (2, 1) = (-3, 2).$$

In general, if $\vec{u} = (a_1, a_2)$ and $\vec{v} = (b_1, b_2)$, then

$$\vec{v} - \vec{u} = (b_1, b_2) - (a_1, a_2) = (b_1 - a_1, b_2 - a_2). \tag{15.5}$$

Remark: Notice in Fig. 15.3b that $\vec{v}_2 = \vec{x} + \vec{v}_1$. (Why?) So it follows that $\vec{x} = \vec{v}_2 - \vec{v}_1$, which justifies its designation.

15.4.3 The zero vector

If $\vec{v} = \vec{u}$ in Eq. (15.5), then $\vec{v} - \vec{u} = (0, 0) = \vec{0}$, the zero vector denoted by $\vec{0}$. When the zero vector is added to any vector it results in the same vector, i.e., $\vec{v} + \vec{0} = \vec{v}$.

15.5. Triangles, Parallelograms and Vectors

15.5.1 Construction of parallelograms and triangles on a geoboard

Consider vectors $\vec{v}_1$ and $\vec{v}_2$ from Fig. 15.1 redrawn in Fig. 15.4. If these vectors are moved parallel to themselves, they form the parallelogram *ABCD*. The parallelogram has two diagonals, *AC* and *BD*. Both *AC* and *BD* can be expressed as vectors: $\overrightarrow{AC} = (1, 4)$ while $\overrightarrow{BD} = (-3, 2)$. We see from Sec. 15.4 that $\overrightarrow{AC} = \vec{v}_1 + \vec{v}_2$ and $\overrightarrow{BD} = \vec{v}_2 - \vec{v}_1$. In addition to parallelogram *ABCD*, vectors $\vec{v}_1$ and $\vec{v}_2$ define ΔABC and ΔABD.

15.5.2 Areas of parallelograms and triangles on a geoboard using determinants

The value of a 2×2 determinant is defined as,

$$\begin{vmatrix} a & b \\ c & d \end{vmatrix} = ad - bc. \tag{15.6}$$

It can be shown that the area of parallelogram *ABCD* in Fig. 15.4 is the absolute value of the 2×2 determinant in which the first row of the determinant is $\vec{v}_1$ while the second row is $\vec{v}_2$:

$$\begin{vmatrix} 2 & 1 \\ -1 & 3 \end{vmatrix} = (2)(3) - (-1)(1) = 7. \tag{15.7a}$$

You can check the result in Eq. (15.7a) by using Pick's Law.

Since ΔABC and ΔABD are half of parallelogram *ABCD*, their areas are each one-half the area of the *ABCD*, i.e.,

$$\text{Area of } \Delta ABC = \frac{7}{2} = \text{Area of } \Delta ABD. \tag{15.7b}$$

Since any polygon can be subdivided into triangles, the area of a polygon can be determined as the sum of the areas of its subdivided triangles as shown in Fig. 15.5. Now we can use vectors and determinants to find the area of these triangles.

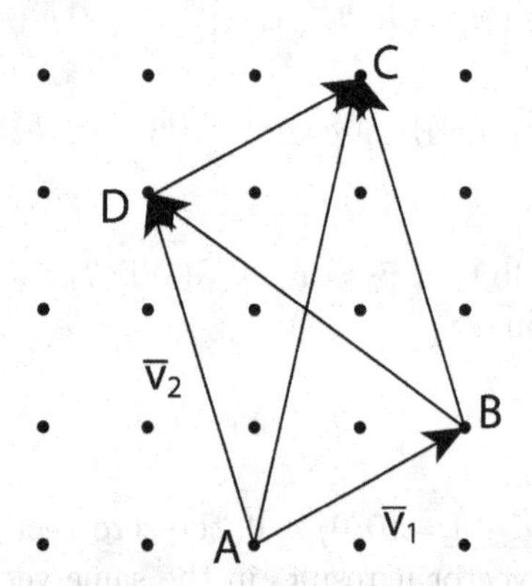

Fig. 15.4. A parallelogram subdivided into two triangles showing its sides and diagonals as vectors.

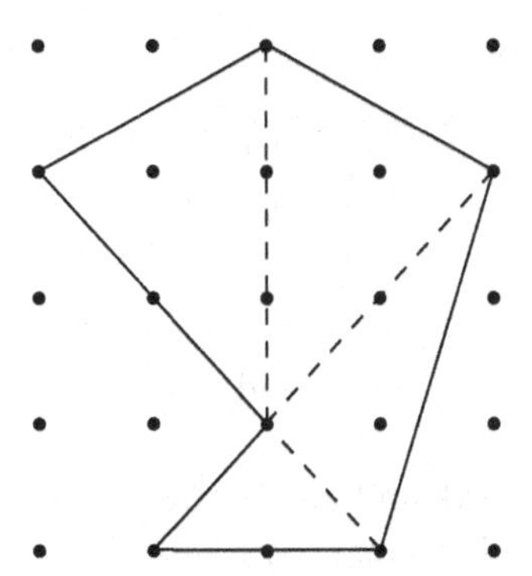

Fig. 15.5. A polygon subdivided into triangles.

15.5.3 Computation of altitudes on a geoboard

The altitude to a side of a triangle can be determined from the area of a triangle,

$$A = \frac{1}{2}bh$$

where b is the length of a side and h is the altitude to that side of the triangle. Solving for the altitude h,

$$h = \frac{2A}{b}. \tag{15.8}$$

The area A can be computed with determinants while b is the length of the side of the triangle to which the altitude is drawn.

Example: Find the altitude in Fig. 15.4, redrawn in Fig 15.6 of ΔABC, to the side given by vector $\overrightarrow{AC} = (1, 4)$. From Eqs. (15.3) and (15.7b),

$$b = \sqrt{1^2 + 4^2} = \sqrt{17} \quad \text{and} \quad A = \frac{7}{2}$$

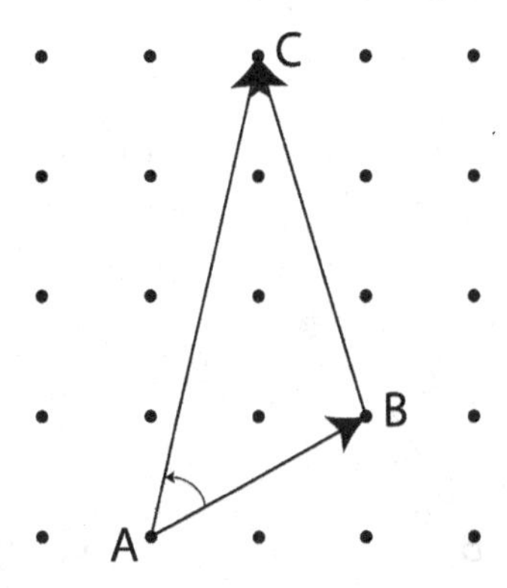

Fig. 15.6. ΔABC from Fig. 15.4 is highlighted.

from Eq. (15.8),

$$h = \frac{7}{\sqrt{17}}.$$

15.6. Problems

1. Create a parallelogram on a geoboard, and find its area by using vectors and determinants.
2. Create a triangle on a geoboard, and find its area by using vectors and determinants. Find the length of an altitude to one of its sides.
3. Create a polygon on a geoboard. Triangulate it and find its area by summing up the areas of the triangles using the determinant method.
4. Find the altitudes in Fig. 15.6 to the sides AB and BC of ΔABC.

15.7. The Law of Cosines

The Law of Cosines can be proven with the help of vectors (see Appendix 15A). In this section, we take another look at the Law of Cosines expressed in terms of vectors. First refer to Fig. 15.7 and restate the Law of Cosines:

$$c^2 = a^2 + b^2 - 2ab\cos C. \tag{15.9}$$

Solving Eq. (15.9) for ab cos C,

$$ab\cos C = \frac{c^2 - a^2 - b^2}{2}. \tag{15.10}$$

Notice that arrows are attached to the sides of triangle ABC so that if,

$$\vec{a} = \overrightarrow{CB} = (a_1, a_2), \quad \vec{b} = \overrightarrow{CA} = (b_1, b_2), \quad \textit{then}$$
$$\vec{c} = \overrightarrow{CB} - \overrightarrow{CA} = (a_1 - b_1, a_2 - b_2) = \overrightarrow{AB},$$

where, $a = |\overrightarrow{CB}|$, $b = |\overrightarrow{CA}|$, $c = |\overrightarrow{CB} - \overrightarrow{CA}|$.

Therefore,

$$a^2 = (a_1^2 + a_2^2), b^2 = (b_1^2 + b_2^2) \quad \text{and} \quad c^2 = (a_1 - b_1)^2 + (a_2 - b_2)^2. \tag{15.11}$$

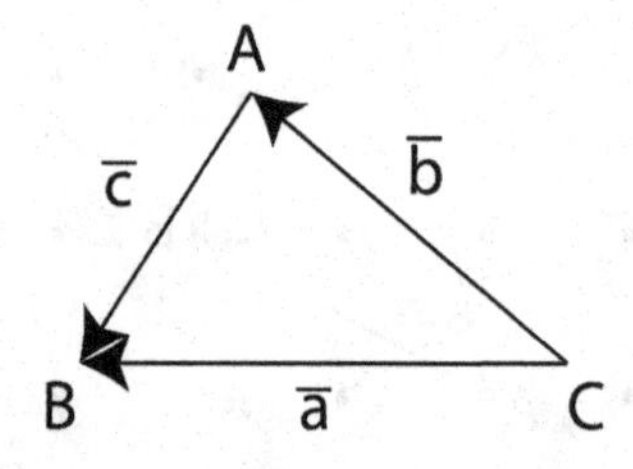

Fig. 15.7. The reference triangle is applied to the Law of Cosines.

Inserting Eq. (15.11) in Eq. (15.10), after some algebra we arrive at the result,

$$ab \cos C = a_1 b_1 + a_2 b_2.$$

Because of its simplicity, ab cos$C$ is denoted by the symbol $\vec{a} \bullet \vec{b}$ called the *dot product* of vectors $\vec{a}$ and $\vec{b}$ i.e.,

$$\vec{a} \bullet \vec{b} = |\vec{a}||\vec{b}| \cos C \tag{15.12}$$

where

$$\vec{a} \bullet \vec{b} = a_1 b_1 + a_2 b_2. \tag{15.13}$$

So that from Eq. (15.12),

$$\cos C = (\vec{a} \bullet \vec{b})/(|\vec{a}||\vec{b}|). \tag{15.14}$$

This will enable us to compute angles between two edges of a triangle with great ease. To find the angle between two vectors,

a. Place their tails together.
b. Compute their lengths using Eq. (15.3).
c. Find their dot product using Eq. (15.13).
d. Compute the cosine of the angle between them using Eq. (15.14).
e. Use a calculator to find the inverse cosine.

Example: Find $\angle A$ between $\vec{v}_1 = (2, 1)$ and $\vec{v}_2 = (1, 4)$ shown in Fig. 15.6.

$$\begin{aligned} |\vec{v}_1| &= \sqrt{5}, \quad |\vec{v}_2| = \sqrt{17} \\ \vec{v}_1 \bullet \vec{v}_2 &= (2)(1) + (1)(4) = 6 \\ \cos C &= \frac{6}{\sqrt{5}\sqrt{17}} = 0.6508 \\ \angle C &= 49.40 \text{ deg.} \end{aligned}$$

15.8. Perpendicular Vectors

If two vectors are *perpendicular*, the angle between them is 90 deg. Since cos 90 = 0, from Eq. (15.12), it is clear that the dot product vanishes. As a result, we conclude:

Two vectors are perpendicular if and only if their dot product is 0.

Example: $(1, 2) \perp (-2, 1)$ and $(2, 2) \perp (-2, 2)$ as shown in Fig. 15.8a.

Problem: As a result of the previous example, these two pair of vectors generate the two squares shown in Figs. 15.8b and 15.8c. These are two of the non-congruent squares that you were asked to find on the geoboard in Chap. 14. With these two squares as examples, can you now find all 8 non-congruent squares that exist on the 5×5 geoboard?

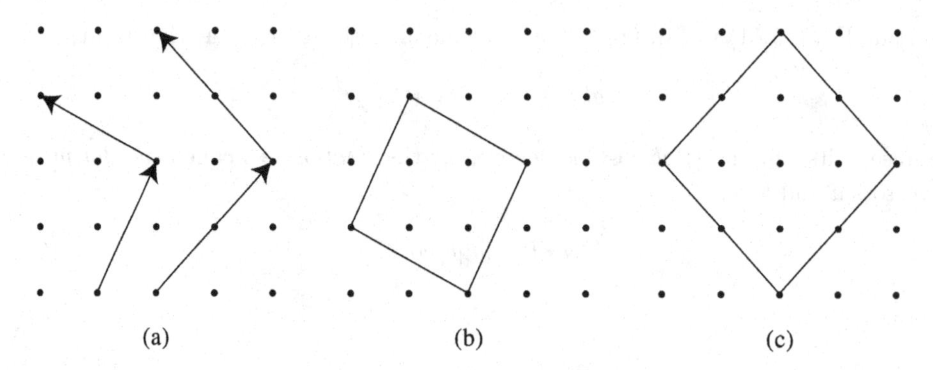

Fig. 15.8. (a) Two pair of perpendicular vectors; (b), (c) the vectors pairs create squares.

15.9. Invariance of the Length of a Vector and the Angle between Vectors Under Rotation

We show that although a pair of vectors change their notation under rotation by 90 deg., their lengths and the angle between them do not change. Therefore, these vectors are *invariant* under 90 deg. rotations. I illustrate this with the following example:

Example:

Vector $\vec{v} = (2, 3)$ is shown on a 5×5 geoboard in Fig. 5.9. Its length is $|\vec{v}| = \sqrt{13}$. Now rotate the geoboard (or your graph paper) 90 deg. in a counterclockwise direction. Notice that the vector has been transformed to vector, $\vec{v} = (-3, 2)$, but its length remains at $|\vec{v}| = \sqrt{13}$.

Remark: It can be shown that the length of the vector is invariant under any rotation.

We also show that the angle between two vectors is also invariant under rotations. This is illustrated by the following example.

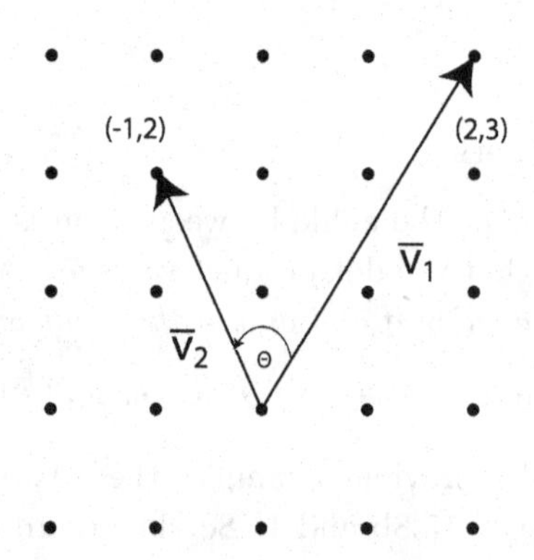

Fig. 15.9. Finding the angle between two vectors.

We compute in Fig. 15.9 the angle between vectors,

$$\vec{v}_1 = (2,3) \quad \text{and} \quad \vec{v}_2 = (-1,2)$$
$$|\vec{v}_1| = \sqrt{13}, \quad |\vec{v}_2| = \sqrt{5}$$
$$\vec{v}_1 \bullet \vec{v}_2 = (2)(-1) + (3)(2) = 4$$
$$\cos\theta = \frac{4}{\sqrt{5}\sqrt{13}} = 0.4961$$
$$\angle\theta = 60.26 \text{ deg.}$$

After a rotation by 90 deg. counterclockwise,

$$\vec{v}_1 = (-3,2) \quad \text{and} \quad \vec{v}_2 = (-2,-1)$$
$$|\vec{v}| = \sqrt{13}, \quad |\vec{v}_2| = \sqrt{5}$$
$$\vec{v}_1 \bullet \vec{v}_2 = (-3)(-2) + (2)(-1) = 4$$
$$\cos\theta = \frac{4}{\sqrt{5}\sqrt{13}} = 0.4961$$
$$\angle\theta = 60.26 \text{ deg.}$$

Problems:

5. Create a triangle on the geoboard and use vectors to compute its angles. Show that the angles sum to 180 deg.
6. Use vectors to express the edge and diagonal of a square as vectors, and then show that the angle between an edge and the diagonal of a square is 45 deg.

15.10. Another Way to Name Vectors

Another way to name a vector is to place it into an (x, y)-coordinate system with the tail of the vector at the origin, (0, 0). The coordinates of the point in the coordinate system which the tip of the vector intersects is then the name of the vector. For example, in Fig. 15.10, vectors $\vec{v}_1$ and $\vec{v}_2$ from Fig. 15.1 are drawn with their tails at the origin. Their tips are at (2, 1) and (–1, 3) respectively.

Remark: With this way of naming vectors, the coordinates of the vector need no longer be integers.

Example: Consider an equilateral triangle with a side of 2 units as shown in Fig. 15.11.

By the Pythagorean Theorem, the altitude is $\sqrt{3}$. The base is placed along the x-axis and two sides of the triangle are denoted by $\vec{v}_1$ and $\vec{v}_2$ where,

$$\vec{v}_1 = (2,0) \quad \text{and} \quad \vec{v}_2 = (1,\sqrt{3}).$$

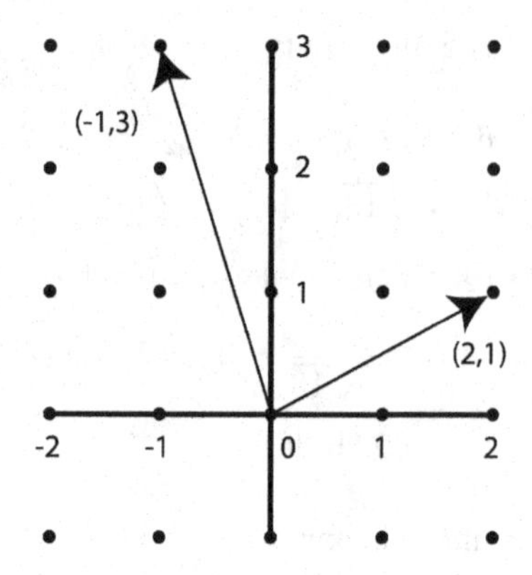

Fig. 15.10. Going beyond integer coordinates, another way to name vectors.

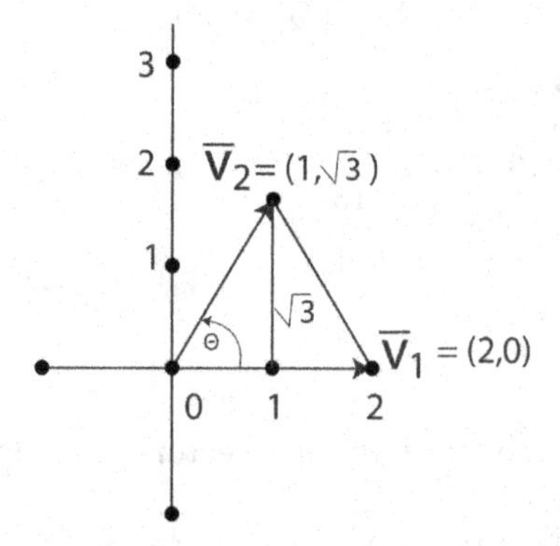

Fig. 15.11. Finding the interior angle of an equilateral triangle using vectors.

We can now verify that the angle between two sides of an equilateral triangle is 60 deg.

$$|\vec{v}_1| = 2$$

$$|\vec{v}_2| = \sqrt{1+3} = 2$$

$$\cos\theta = (\vec{v}_1 \bullet \vec{v}_2)/(|\vec{v}_1||\vec{v}_2|) = \frac{(2,0) \bullet (1,\sqrt{3})}{4} = \frac{2}{4} = \frac{1}{2}.$$

Therefore,

$$\theta = 60 \text{ deg.}$$

Appendix 15A. Proof of the Law of Cosines

Theorem (Law of Cosines): For ΔABC, $c^2 = a^2 + b^2 - 2ab\cos\theta$.

Proof:

Consider ΔABC where vertex C is at the origin of an (x, y)-coordinate system as shown in Fig. 15A.1. Point A is represented by vector $\vec{a} = (a\cos\theta_1, a\sin\theta_1)$, point B is represented by vector $\vec{b} = (b\cos\theta_2, b\sin\theta_2)$ and θ is the angle opposite side c. Side c is then represented by the vector from the tip of $\vec{a}$ to the tip of $\vec{b}$ or $\vec{b} - \vec{a}$.

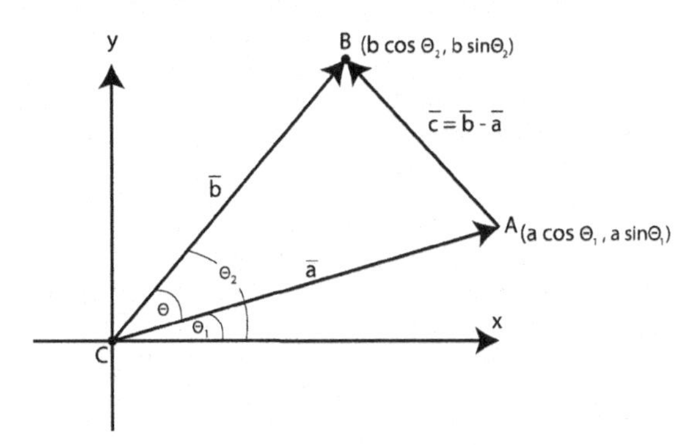

Fig. 15A.1. Proof of the law of cosines.

After doing some algebra, the Law of cosines is simply the statement that,

$$c = |\vec{b} - \vec{a}| \tag{15A.1}$$

where $|\vec{a}| = a$, $|\vec{b}| = b$. $|\vec{c}| = c$.

Here are the details of the algebra:

$$\vec{b} - \vec{a} = (b\cos\theta_2 - a\cos\theta_1, b\sin\theta_2 - a\sin\theta_1).$$

Therefore,

$$\begin{aligned} c^2 &= |\vec{b} - \vec{a}|^2 = (b\cos\theta_2 - a\cos\theta_1)^2 + (b\sin\theta_2 - a\sin\theta_1)^2 \\ &= b^2(\cos^2\theta_2 + \sin^2\theta_2)^2 + a^2(\cos^2\theta_1 + \sin^2\theta_1) - 2ab(\cos\theta_2\cos\theta_1 + \sin\theta_2\sin\theta_1) \end{aligned}$$

Since $(\cos^2\theta_1 + \sin^2\theta_1 = 1 = \cos^2\theta_2 + \sin^2\theta_2)$, and an identity from trigonometry states,

$$(\cos\theta_2\cos\theta_1 + \sin\theta_2\sin\theta_1) = \cos(\theta_2 - \theta_1),$$

it follows that,

$$c^2 = a^2 + b^2 - 2ab\cos\theta.$$

where $(\theta_2 - \theta_1) = \theta$. QED

CHAPTER 16

LOGARITHMIC SPIRALS

16.1. Introduction

The *logarithmic spiral* can be thought of as the curve of life in that it is self-symmetric and occurs in the shape of sea animals called Nautilus shells, the striations of sea shells, and the horns of horned animals (see Fig. 16.1). A logarithmic spiral is formed when a horn grows faster on the outside than the inside, illustrated in Fig. 16.1 with wooden blocks cut by a perpendicular plane. If the plane cuts the block at an angle, the growth pattern is helical.

That the logarithmic spiral is *self-similar* at all scales follows from its property that any segment of the spiral that subtends an angle θ is similar to any other segment subtending the same angle as shown in Fig. 16.2. In other words, if one such arc of the spiral is magnified or contracted it lies on top of the others. Therefore, any scaling maps the logarithmic spiral onto itself. The log spiral is the only smooth curve with

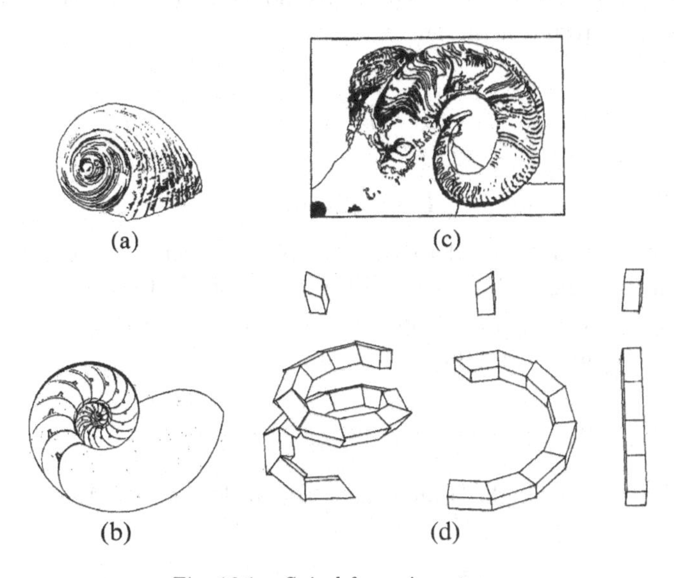

Fig. 16.1. Spiral forms in nature.

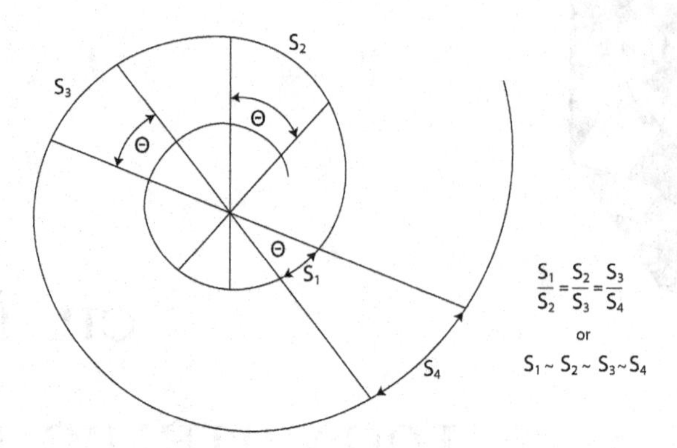

Fig. 16.2. Self-similarity of logarithmic spirals.

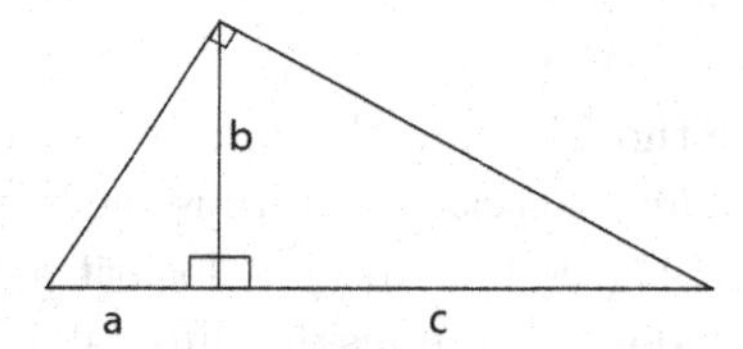

Fig. 16.3. Three similar right triangles.

this property. In Chap. 23, we will see that a class of non-smooth curves known as fractals also share this property.

The logarithmic spiral is generated by a right triangle making use of its self-similarity that we discovered by performing surgery on a right triangle in Chap. 12. We labeled the edges meeting at the intersection of the hypotenuse and the altitude from the opposite vertex by the letters a, b, c as shown in Fig. 12.6, redrawn in Fig. 16.3, and we showed, as a result of similarity, that,

$$\frac{a}{b} = \frac{b}{c}. \tag{16.1}$$

16.2. From Right Triangle to Logarithmic Spiral

How does the right triangle lead to the log spiral? In an (x, y)-coordinate system, construct a right triangle with base 1 and height k units and continue this to a second right triangle as shown in Fig. 16.4a. Do you notice the same configuration as in Fig. 16.3? Therefore from Eq. (16.1),

$$\frac{1}{k} = \frac{k}{x} \quad \text{or} \quad x = k^2.$$

Continue this to the third right triangle as shown in Fig. 16.4b. Again,

$$\frac{k}{k^2} = \frac{k^2}{x} \quad \text{or} \quad x = k^3.$$

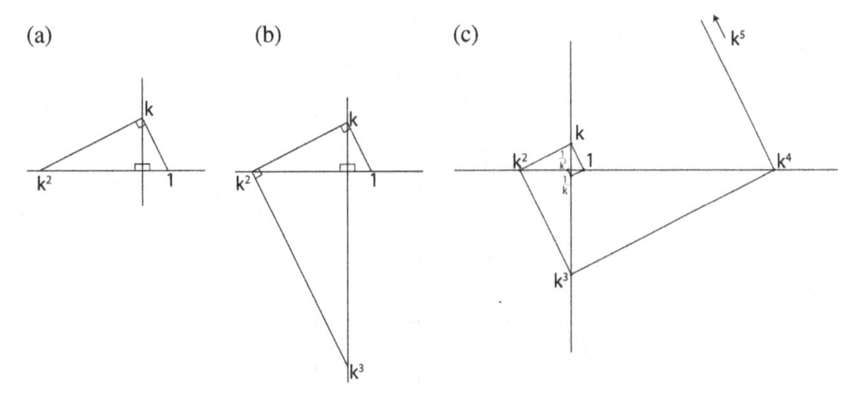

Fig. 16.4. The vertex points on a log spiral (a) 3 vertex points; (b) expanded to 4 vertex points; (c) inwardly and outwardly spiraling vertex points.

Continuing in this manner we see that we are generating a sequence of *vertex points* of an outwardly spiralling form known as a *logarithmic spiral.* The values $1, k, k^2, k^3, k^4, \ldots$ are the *radii* of the spiral at these vertex points, and they form a *geometric sequence* of vertices shown in Fig. 16.4b.

Note that we can also spiral inwardly where, again from Eq. (16.1),

$$\frac{k}{1} = \frac{1}{x} \quad \text{or} \quad x = \frac{1}{k}$$

as shown in Fig. 16.4c. So the radii of a double spiral form the double geometric sequence:

$$\cdots \frac{1}{k^3}, \frac{1}{k^2}, \frac{1}{k}, 1, k, k^2, k^3, k^4, \ldots. \tag{16.3a}$$

If $k = 2$ we get the double geometric sequence:

$$\cdots \frac{1}{2^2}, \frac{1}{2}, 1, 2, 2^2, 2^3, 2^4, \ldots. \tag{16.3b}$$

These results are recorded in Table 16.1 and illustrated in Fig. 16.4c for the case of $k = 2$. The left-hand column of this table records the number n of right angles, 90 deg., of the defined vertices while the right column is the distance r from the center of the spiral located at the origin to each vertex. Angles are measured from the 0-deg. ray in the direction of the positive x-axis. Positive angles are in a counterclockwise direction and negative angles are in a clockwise direction from the 0-deg. ray, similar to the convention we used in Chap. 6.

This represents the table of an *exponential function* base 2, i.e., $y = 2^x$, which is defined for all real numbers x, shown as a graph in Fig. 16.5. Reading Table 16.1 backwards, you get the table for a *logarithm* base 2 or, $y = \log_2 x$, e.g., $\log_2 1 = 0$, $\log_2 2 = 1$, $\log_2 4 = 2$, etc. For this reason the spiral is called a *logarithmic spiral.* The

Table 16.1. Vertex Points on a Logarithmic Spiral

$n = \frac{\theta}{90}$	r
-2	$\frac{1}{4}$
-1	$\frac{1}{2}$
0	1
1	2
2	4
3	8
4	16
5	32
6	64
n	2^n

graph of $y = \log_2 x$ is also shown in Fig. 16.5. In Table 16.1, $k = 2$. If k is taken to be a different value, you would get a different geometric sequence and logarithms to a different base. Notice that whatever the base, $\log_k 1 = 0$. Also the log of any number to its own base is 1.

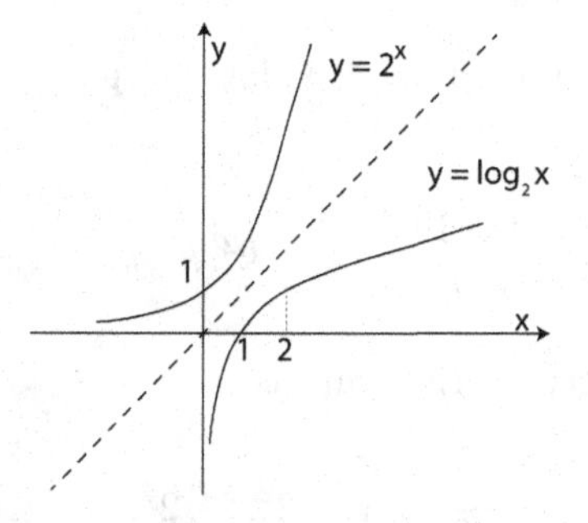

Fig. 16.5. Graphs of $y = 2^x$ and $y = \log_2 x$.

An important property of logarithms is evident from Table 16.1. In Column 1, you will see that $2 + 3 = 5$ correlates in Column 2 with $4 \times 8 = 32$. Also $5 - 3 = 2$ correlates with $32 \div 8 = 4$. In other words, when a pair of numbers multiply or divide their logarithms add or subtract. This remarkable property of logarithms enabled numbers with many digits to be multiplied (or divided) by simply adding (or subtracting) their logarithms. Since addition is so much easier to carry out than multiplication (or division), before the advent of the computer or the hand calculator, this was the only way to do these operations, provided you had a sufficiently extensive log table handy. Later, slide rules were invented that enabled 3- or 4-digit multiplications or divisions to be carried out mechanically.

Another important property of logarithms can be seen by noticing from Table 16.1 that when 4 cubes to $4^3 = 64$, log 4 triples from 2 to 6. In other words, when a number is taken to a power its logarithm gets multiplied by that power.

The properties of logarithms can be summarized as follows:

$$\log ab = \log a + \log b$$
$$\log \frac{a}{b} = \log a - \log b$$
$$\log a^b = b \log a$$
$$\log 1 = 0$$
$$\log_k k = 1$$

Remark: When the base of logarithms is not mentioned the relationship works in any base.

Also notice from Table 16.1 the following principle of growth for log spirals:

***Constructive Principle*:** If you double the angle of a log spiral, its radius squares.

From this property of log spirals, one is able to construct a dense set of points on the spiral using only compass and straightedge. The next section presents a simple method for doing this.

16.3. A Simple Way to Generate a Dense Set of Points on a Logarithmic Spiral

Look at the geometric Sequence (16.3b) rewritten as Sequence (16.4).

$$\cdots \frac{1}{4} \; \frac{1}{2} \; 1 \; 2 \; 4 \; 8 \; 16 \; 32 \; 64 \cdots \tag{16.4}$$

Do you notice that if you take any three consecutive numbers from this sequence, the middle number is the square root of the product of the other two, e.g., for $2, 4, 8$: $\sqrt{2 \times 8} = 4$, for $4, 8, 16$: $\sqrt{4 \times 16} = 8$, etc. The middle number is called the geometric mean of the other two. In general, given two numbers a and b we can define their mean, c in many ways. You are most familiar with the arithmetical mean or average $c = \frac{a+b}{2}$. However, the geometric mean is $c = \sqrt{a \times b}$.

The elements of Sequence (16.4) are the radii of the logarithmic spiral that appear in Table 16.1 at angles: $\frac{\theta}{90} = \cdots -2 \;\; -1 \;\; 0 \;\; 1 \;\; 2 \;\; 3 \;\; 4 \cdots$. I refer to these as the *vertex points* on the spiral. Points intermediate to these on the spiral can be found by choosing consecutive pairs of radii and from Table 16.1 and determining their geometric means which occur at angles midway between the corresponding pair of angles, e.g., for

r: 1 2 occurring at $\frac{\theta}{90}$: 0 1, the geometric mean is $\sqrt{2}$ which occurs at $\frac{\theta}{90} = \frac{1}{2}$ (45 deg.)

r: 2 4 occurring at $\frac{\theta}{90}$: 1 2, the geometric mean is $\sqrt{8} = 2\sqrt{2}$ which occurs at $\frac{\theta}{90} = \frac{3}{2}$ (135 deg.)

Table 16.2. Additional Points on a Logarithmic Spiral

$\frac{\theta}{90}$	r
$\frac{-1}{2}$	$\frac{1}{\sqrt{2}}$
$\frac{1}{2}$	$\sqrt{2}$
$\frac{3}{2}$	$2\sqrt{2}$
$\frac{5}{2}$	$4\sqrt{2}$
$\frac{7}{2}$	$8\sqrt{2}$
...	

r: 4 8 occurring at $\frac{\theta}{90}$: 2 3, the geometric mean is $\sqrt{32} = 4\sqrt{2}$ which occurs at $\frac{\theta}{90} = \frac{5}{2}$ (225 deg.)

Continuing in this fashion results in Table 16.2 which is an extension of Table 16.1.

Do you see a pattern to the radii? In a second iteration, the geometric means of the new intervals that have been formed are computed, e.g., for,

r : 1 $\sqrt{2}$ occurring at $\frac{\theta}{90}$: 0 $\frac{1}{2}$, the geometric mean is $\sqrt{\sqrt{2}}$ which occurs at $\frac{\theta}{90} = \frac{1}{4}$ (22.5 deg.)

r : $\sqrt{2}$ 2 occurring at $\frac{\theta}{90} = \frac{1}{2}$ 1, the geometric mean is $\sqrt{2\sqrt{2}}$ which occurs at $n = \frac{3}{4}\ldots$ (67.5 deg.)

In this way a dense set of points from the logarithmic spiral can be determined. And since they require only square roots, the method of Sec. 13.6 can be used to construct them with compass and straightedge.

Problem: Using the procedure above, construct six additional points on the logarithmic spiral in addition to the vertex points.

16.4. The Law of Repetition of Ratios

The logarithmic spiral is also intrinsic to a construction used to replicate proportions of a rectangle known as the *Law of Repetition of Ratios.* It was used during the classical era in ancient Greek and Roman architecture and design, and later in the Renaissance. Its importance was recognized by Jay Hambridge [Ham], [EdwE] who coined the term *dynamic symmetry* to describe this process.

16.4.1 Dynamic symmetry

Begin with some geometric form or pattern which we call a *unit* and add another form or pattern called a *gnomon* G which enlarges the unit U while preserving its shape. In this sense, consider the unit to be rectangle ABCD in Fig. 6.6a with sides in the ratio $a : b$ and draw a line segment EC from one vertex that intersects at right angles at O a diagonal DB of the rectangle as shown in Fig. 16.6a. Then draw EF perpendicular to AB with F on DC. In this way the rectangle is divided into two rectangles one of which is a unit with the same proportions $a : b = b : c$ at a smaller scale and the other being the gnomon, i.e., U = U + G as shown in Fig. 16.6b. We see here that this construction creates three similar right triangles ΔDOC, ΔCOB, ΔBOE for which,

$$\frac{a}{b} = \frac{b}{c}, \tag{16.5}$$

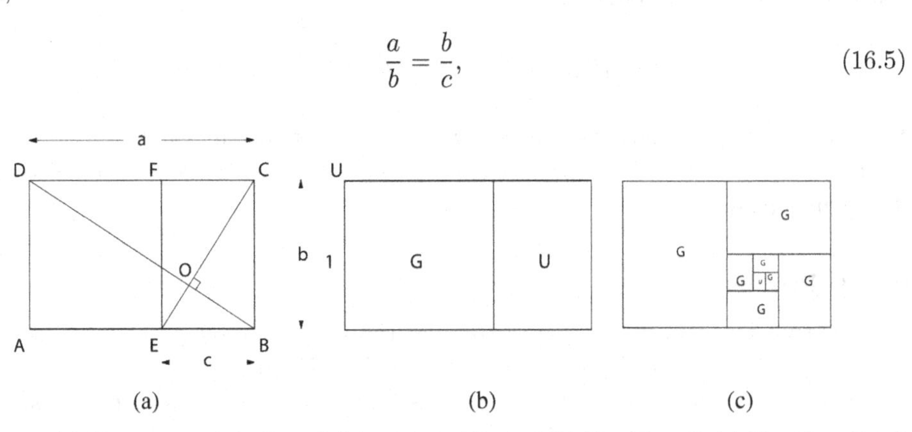

Fig. 16.6. (a) Illustration of the Law of Repetition of Ratios; (b) U = U + G; (c) U = G + G +G $+ \cdots +$ U (for the case where G = U).

where this time a, b, c are the hypotenuses of these three similar triangles.

Looking again at Fig. 16.6a you will see that the same construction is ready to be carried out at a smaller scale so that U = U + G + G. This process can be continued indefinitely so that the unit is tiled by a sequence of whirling gnomons and one final unit, U = G + G + G + ... G + U as shown in Fig. 16.6c for the case where $U = G$.

16.4.2 The gnomon is a square

If the gnomon is a square, i.e., G = S, what is the proportions of the unit (see Fig. 16.7a)? Let the proportions of U be $x : 1$, in which case: $a = x, b = 1, c = x - 1$.

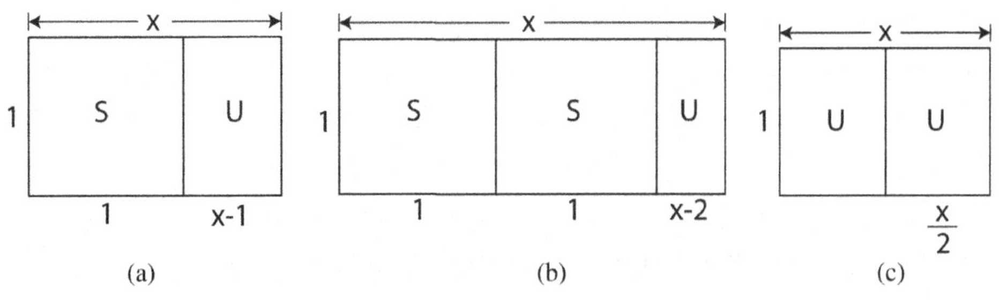

Fig. 16.7. The Law of Repetition of Ratios applied to units whose gnomon is a a)square; b) double square; c) the unit.

Then using Eq. (16.5),

$$\frac{x}{1} = \frac{1}{x-1}$$

or

$$x^2 - x - 1 = 0. \tag{16.6}$$

Using the quadratic formula (see Sec. 6.4)

$$x = \frac{1+\sqrt{5}}{2} = \phi, \quad \text{(Verify this!)}$$

A number emerges known as the *golden mean* with many unusual properties. This number will be discussed in the next chapter and be shown to lead to a construction of a regular pentagon.

16.4.3 Problems

a. Show that when G = DS (double square), the proportions $x : 1$ of the unit is $x = 1 + \sqrt{2} = \theta$ known of as the silver mean (see Fig. 16.7b).
b. When G = U, then $x = \sqrt{2}$. In other words, when a rectangle has proportions: $\sqrt{2} : 1$ (root 2 rectangle), and it is divided in half, it yields a pair of root-2 rectangles (see Fig. 16.7c). This is also the unit that was illustrated in Fig. 16.6c.

The silver mean governs a geometry based on $\sqrt{2}$ and octagons and will be discussed in the next chapter.

16.5. Whirling Squares

Since the gnomon of the golden rectangle is a square, quarter circle arcs can be drawn within the sequence of *whirling squares* to form an excellent approximation to a golden logarithmic spiral as shown in Fig. 16.8.

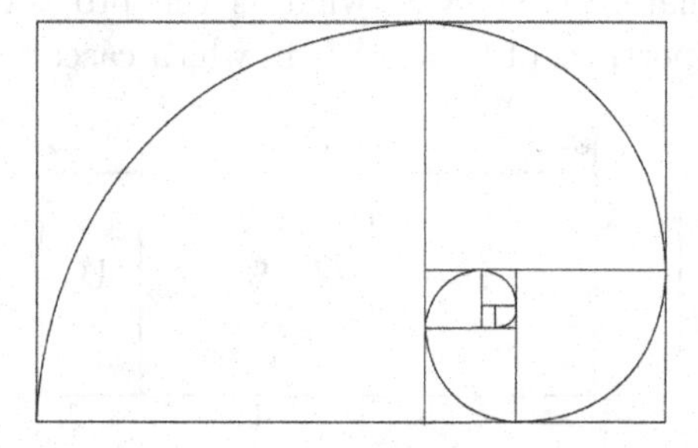

Fig. 16.8. Approximate construction of a golden spiral.

Construction 1: Use the whirling squares to construct an approximate logarithmic spiral.

16.6. Additional Constructions

Construction 2: Follow the instructions in Appendix 16A to draw the four logarithmic spirals traversed by Tom Pizza's trained turtles as described by Martin Gardner [Gar].

Construction 3: Follow the constructions in Appendix B to construct a Baravelle spiral.

Appendix 16A. The Four Turtle Problem

Tom Pizza has trained his four turtles so that Abner always crawls toward Bertha, Bertha toward Charles, Charles toward Delilah, and Delilah toward Abner. One day he put the four turtles in this order, $ABCD$ at the four corners of a square room. He and his parents watched to see what would happen.

"Very interesting, Son," said Mr. Pizza. "Each turtle is crawling directly toward the turtle on its right. They all go at the same speed, so at every instant, they are at the corners of a square." (See Fig. 16A.1.)

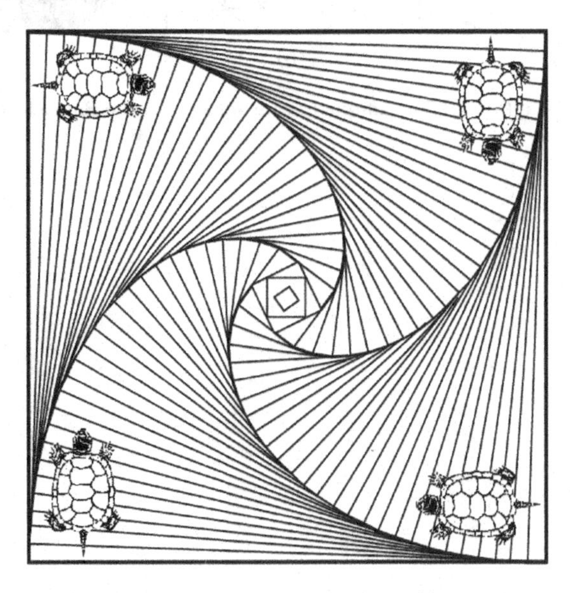

Fig. 16A.1. The four turtle problem.

"Yes, Dad" said Tom, "and the square keeps turning as it gets smaller and smaller. Look! They're meeting right at the center."

Assume that each turtle crawls at a constant rate of 1 centimeter per second and that the square room is 3 meters on the side. How long will it take the turtles to meet at the center? Of course, we must idealize the problem by thinking of the turtles as points.

Mr. Pizza tried to solve the problem by calculus. Suddenly Mrs. Pizza shouted, "You don't need calculus, Pepperone! It's simple. The time is 5 minutes."

What was Mrs. Pizza's insight? If you cannot provide the requisite insight to solve this problem, you can always diagram the paths of the turtles in small increments of time, drawing four sides of the square at the end of each interval. The result is the pattern shown in Fig. 16A.1

Appendix 16B. The Baravelle Spiral

In Fig. 16B.1a, we see that a circle is drawn tangent to an outer square (inscribed circle) and touching the vertices of an inner square (circumscribed circle). This square-within-a-square, called *an ad-quadratum square*, was much used in ancient geometry. This is continued to additional ad-quadratum squares in Fig. 16B.1b. Regions of Fig. 16B.1b are shaded resulting in the Baravelle spiral in Fig. 16B.1c. A similar construction can be carried out to create Baravelle spirals beginning with any regular polygon.

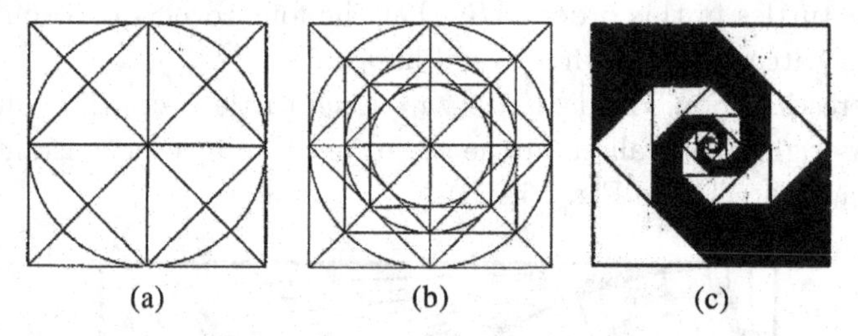

Fig. 16B.1. (a), (b) The ad-quadratum square; (c) the Baravele spiral.

CHAPTER 17

THE GOLDEN AND SILVER MEANS

17.1. Introduction

Fine artists, composers, architects, scientists and engineers have often created their best works by keeping their eyes open to the natural world. The natural world consists of an elegant dialogue between order and chaos. Careful study of a cloud formation or a running stream shows that what at first appear to be random fluctuations in the observed patterns are actually subtle forms of order. Mathematics is the best tool that humans have created to study the order in things.

Despite the infinite diversity of nature, mathematics and science have attempted to reduce this complexity to a few general principles. In this chapter, we introduce you to one enigmatic number, the golden mean, which appears and reappears throughout art and science [Kap1], [Kap2], [Liv], [Ols], [Sta]. We show that this number is actually a member of a family of numbers known as silver means which have applications to architecture and design as well as mathematics and science.

17.2. The Golden Mean and the Golden Section

In the last workshop we encountered a number $\phi = \frac{1+\sqrt{5}}{2}$ known as the golden mean. The story of the golden mean begins with an undifferentiated line segment.

The line segment is divided into two parts such that the ratio of the whole to the largest segment equals the ratio of the larger to the smaller as shown in Fig. 17.1.

From this simple statement all follows about this extraordinary number. This is consistent with Gordon Spencer Brown's notion that once you make a mark on an undifferentiated medium this determines the universe of discourse that unfolds [Spe-B].

First of all, the above statement can be written as the following three equations where x is the larger segment and y the smaller (make sure that you understand where

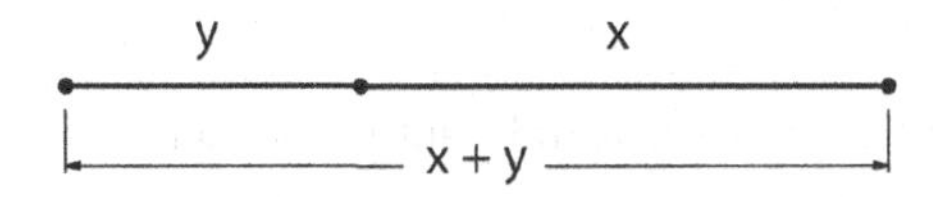

Fig. 17.1. A line segment is sectioned into two parts.

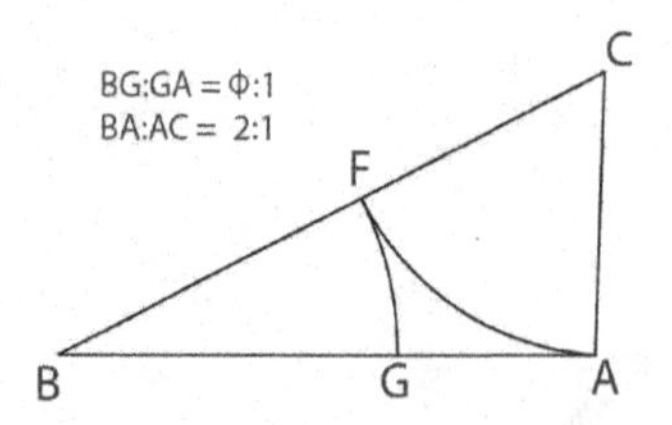

Fig. 17.2. Construction of the golden section.

these equations come from):

$$\frac{x+y}{x} = \frac{x}{y} \tag{17.1}$$

$$z - \frac{1}{z} = 1 \quad \text{where } z = \frac{x}{y} \tag{17.2}$$

$$1 + z = z^2. \tag{17.3}$$

The positive solution to these equations is $z = \phi = \frac{1+\sqrt{5}}{2}$, the *golden mean*, and so the original undifferentiated line segment is divided into a ratio of $1 : \phi$ referred to as the *golden section.* Some time ago, I discovered a wonderful construction of the golden section in the notebooks of the artist Paul Klee. It is displayed in Fig. 17.2 where $AC : AB = 1 : 2$. It can be constructed as follows:

Construction 1: Golden section

To divide a given line segment into two line segments the ratio of whose lengths are the golden mean, i.e., in the golden section, proceed as follows (see Fig. 17.2):

a. Start with a line segment AB.
b. Draw $AC = \frac{1}{2}AB$.
c. Circular arc of radius CA intersects CB at F.
d. Circular arc of radius BF intersects AB at G breaking AB into a golden section.

Equation (17.1) expresses the fact that the golden mean creates a proportion in which the whole is similar to its parts. In Chap. 23, this will be seen to be the principal characteristic of a fractal.

Some would say that the golden mean is "number one" in terms of its importance. This statement can be justified by taking Eq. (17.2) and rewriting it as,

$$\phi = z = 1 + \frac{1}{z} = 1 + \frac{1}{1+\frac{1}{z}} = 1 + \frac{1}{1+\frac{1}{1+\frac{1}{z}}} = \cdots = 1 + \frac{1}{1+\frac{1}{1+\frac{1}{1+\frac{1}{1+\cdots}}}}. \tag{17.4}$$

Cutting off these continued fractions at different levels yields the set of approximations,

$$1, \frac{3}{2} = 1.5, \frac{5}{3} = 1.667, \frac{8}{5} = 1.6, \frac{13}{8} = 1.625, \ldots.$$

These are the ratios of successive terms from the *Fibonacci F-sequence*,

$$1\ 1\ 2\ 3\ 5\ 8\ 13\ 21, \ldots. \tag{17.5}$$

These ratios approximate better and better, in terms of decimal values, the golden mean $\phi = \frac{1+\sqrt{5}}{2}$ which is the irrational number (non-repeating decimal) 1.61803.... We say that ϕ is the *limit* of the sequence of ratios of successive terms of the sequence (17.5). This is illustrated graphically in Fig. 17.3.

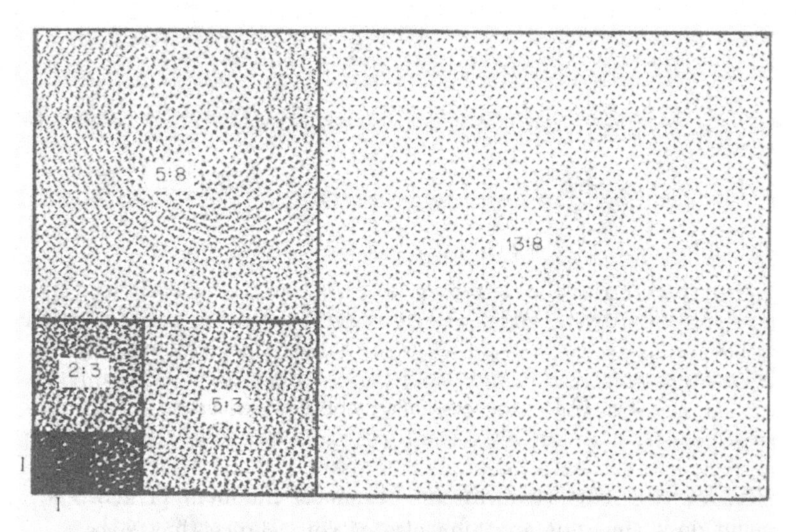

A rectangle of any proportions may be expanded to asymptotically approximate a golden rectangle illustrated for a 1:1 square.

Fig. 17.3. Ilustration of the ratio of successive terms of an F-sequence.

Problem 1: How far out in the Fibonacci sequence 17.5 do you have to go to get agreement with the golden mean to five decimal places?

A Fibonacci sequence is a sequence of integers that satisfies the recursion formula,

$$a_n = a_{n-1} + a_{n-2} \tag{17.6}$$

which states that each integer of the sequence is the sum of the previous two numbers.

This sequence is very much connected with the growth of plants and other natural phenomena [kap1], [kap2]. The additive nature of this sequence led the architect LeCorbusier to use it as the basis of a system of proportions that he called the Modulor [kap1].

In Fig. 17.4, counting the I's at each level results in the F-sequence. But notice that this pattern of I's is repeated over and over at different levels. For this reason, the F-sequence is said to be self-similar.

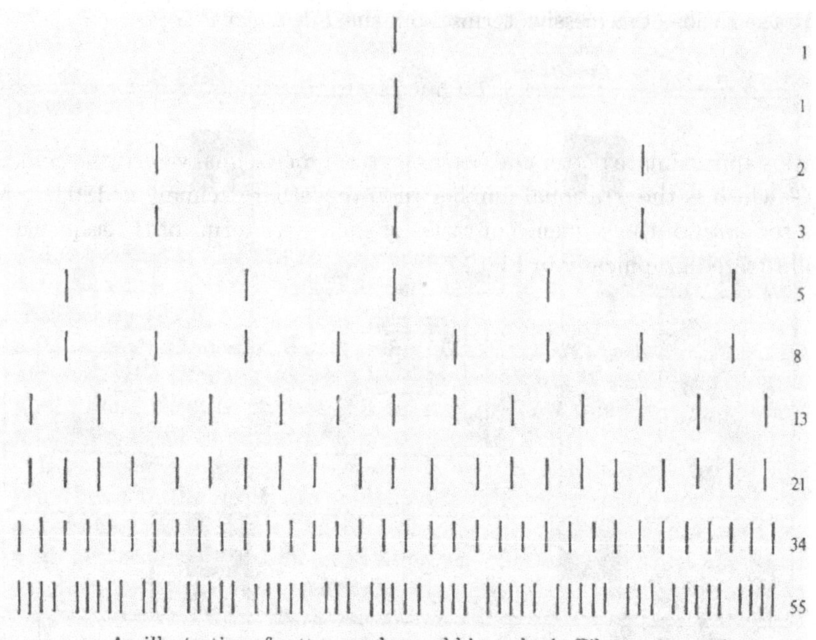

Fig. 17.4. The self-similarity of the F-sequence.

Problem 2: Create a pattern from the Fibonacci numbers $\{1, 2, 3, 5, 8, \ldots\}$. Your pattern can be dots, lines, or anything else of your choice that gives a geometrical rendering of the Fibonacci sequence.

17.3. The Silver Mean

In Chap. 16, we also encountered a second number, $\theta = 1 + \sqrt{2}$, known as the *silver mean* that satisfies the equation,

$$z^2 - 2z - 1 = 0 \tag{17.7}$$

which can be rewritten as,

$$z - \frac{1}{z} = 2.$$

The golden and silver means are the first two numbers of a family of numbers that satisfy the equation,

$$z - \frac{1}{z} = n \tag{17.8}$$

and has interesting mathematical and geometrical properties.

Now consider the silver mean $\theta = 1 + \sqrt{2}$ which satisfies Eq. (17.7).

A similar argument to the golden mean can be made to show that,

$$\theta = 2 + \cfrac{1}{2 + \cfrac{1}{2 + \cfrac{1}{2 + \cfrac{1}{2 + \cdots}}}}. \tag{17.9}$$

Its approximations are the ratio of terms from another sequence known as *Pells Sequence*:

$$1, 2, 5, 12, 29, 70, \ldots \tag{17.10}$$

satisfying,

$$a_n = 2a_{n-1} + a_{n-2}. \tag{17.11}$$

This sequence was used as the basis of the proportional system that created much of the architecture of the Roman Empire [Wat].

The ratio of successive terms from Pells sequence approaches closer and closer to $\theta = 1 + \sqrt{2}$ in the sense of a limit.

Remark: Sequences (17.5) and (17.10) are approximate geometric sequences (see Sec. 16.3) in the sense that given three successive terms from them, the product of the first and third is equal to the square of the middle term off by 1 unit, e.g., from Seq. (17.5), $2 \times 5 = 3^2 + 1$ while from Seq. (17.10), $2 \times 12 = 5^2 - 1$.

17.4. The Golden Spiral

When $k = \phi$, the *golden mean,* this results in a spiral just like the one in Fig. 16.4. The vertex points on the *golden logarithmic spiral* again form a double geometric ϕ-Sequence (17.12):

$$\cdots \frac{1}{\phi^3}, \frac{1}{\phi^2}, \frac{1}{\phi}, 1, \phi, \phi^2, \phi^3, \phi^4 \cdots. \tag{17.12}$$

This geometric sequence has another wonderful property. It is also a Fibonacci sequence, i.e., each number in the sequence is the sum of the two previous numbers. For example

$$\frac{1}{\phi^2} + \frac{1}{\phi} = 1, 1 + \phi = \phi^2, \quad \text{etc.}$$

Problem 3: Show by arithmetic that,

a. $\left(\frac{1+\sqrt{5}}{2}\right)^2 = 1 + \left(\frac{1+\sqrt{5}}{2}\right)$, b. $\frac{1}{\left(\frac{1+\sqrt{5}}{2}\right)^2} + \frac{1}{\left(\frac{1+\sqrt{5}}{2}\right)} = 1$

17.5. A Golden Rectangle

A golden mean rectangle can be constructed with compass and straightedge as follows (see Fig. 17.5):

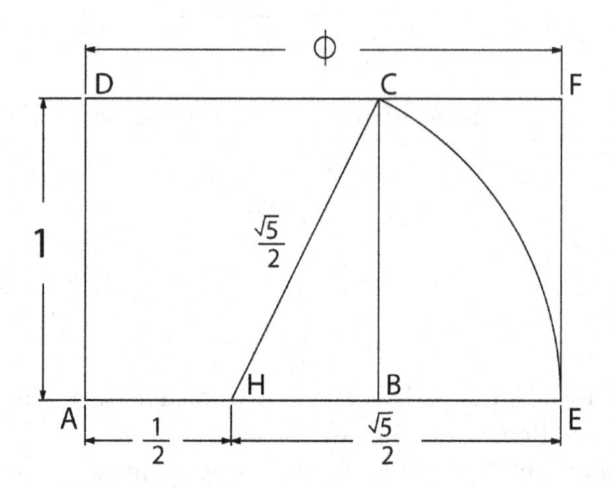

Fig. 17.5. A golden rectangle.

Construction 2: A Golden Rectangle

a. Begin with a square $ABCD$ of side 1 unit.
b. Bisect the base AB.
c. Extend the base.
d. Place your compass point at the midpoint, H of the base and use a compass to sweep out an arc with radius HC until it intersects the extended baseline AB at E.
e. Since $HC = \frac{\sqrt{5}}{2}$, we have that $AE = \phi = \frac{1+\sqrt{5}}{2}$.

Construction 3: The phi sequence
Now that you have lengths 1 and ϕ, you can construct all of the terms in Seq. (17.12) by making use of the Fibonacci property. This is shown in Fig. 17.6.

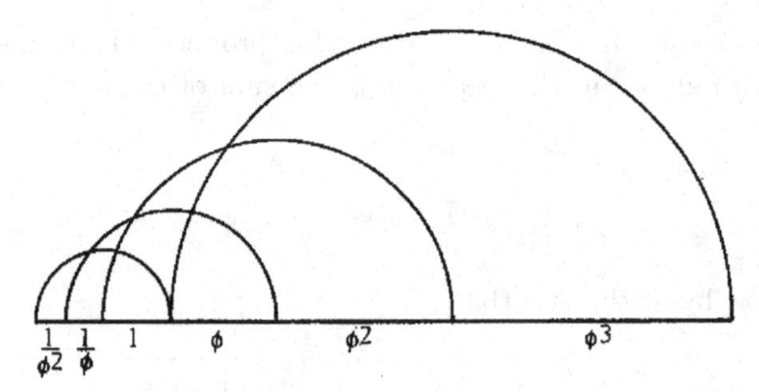

Fig. 17.6. Construction of the ϕ sequence with compass and straightedge beginning with 1 and ϕ.

We have also seen in Sec. 16.4.2 that if you remove a square from a golden rectangle you are left with another golden rectangle at a smaller scale.

Problem 4: Show for Fig. 17.5 that the ratio of sides of the rectangle *BEFC* is $1 : \phi$.

By repeating this action we arrive at a golden rectangle tiled by whirling squares as described in Sec. 16.5 and shown in Fig. 16.8.

17.6. A Regular Pentagon

You can also use the golden mean to construct a *regular pentagon* since the diagonal of a pentagon has the value ϕ when the length of the side is 1 unit as shown in Fig. 17.7. Therefore, the pentagon can be constructed by the following procedure:

Construction 4: A Regular Pentagon

a. Begin with line segment *AB* of length 1 unit where *A* and *B* are two vertices of the pentagon.
b. Sweep out two arcs from *A* and *B* of length ϕ units where 1 and ϕ have been constructed from the golden section as in Sec. 17.5. The two arcs intersect at *C*, another vertex of the pentagon.
c. Sweep out two arcs at *B* and *C* of length 1. Where they intersect is vertex *D*.
d. Do the same at vertices *A* and *C* to form vertex *E* of the pentagon.

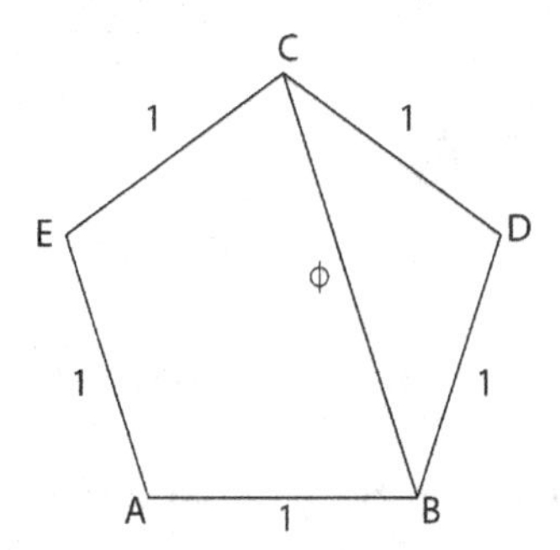

Fig. 17.7. Construction of a pentagon.

17.7. Golden Triangles

From the pentagon, you can construct two isosceles *golden triangles* as shown in Fig. 17.8a (see Sec. 8.5 problem 2b). Triangle 1 has base 1 and sides ϕ with base angles equal to 72 deg. and the remaining vertex angle of 36 deg. Triangle 2 has a base of ϕ and sides of 1 unit. Because of this simple geometry, these triangles are self-similar in the following respect. If you bisect the 72 deg. angle of Triangle 1, it results in similar versions of Triangles 1 and 2 at a smaller scale (see Fig. 17.8b). This construction can be continued over and over to construct triangles at smaller and smaller scales (see Fig. 17.8c).

Construction 4: Construct golden triangles 1 and 2 at three different scales and fit them together into a pleasing design. Two examples of student projects are shown in Fig. 17.9.

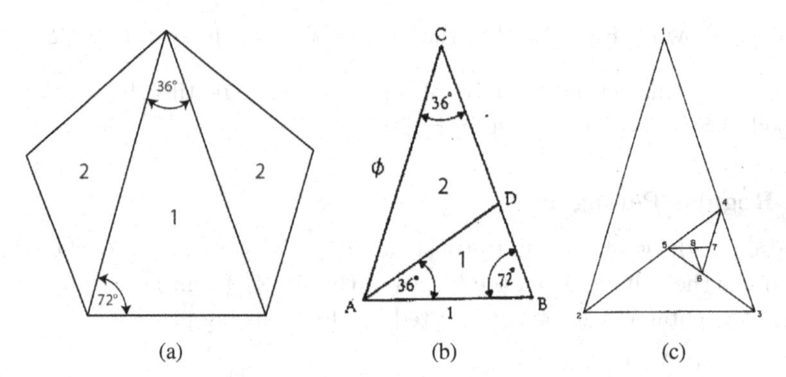

Fig. 17.8. (a) A pentagon subdivided into type 1 and type 2 golden triangles; (b) the base angle of a type 1 golden triangle is bisected to form a type 1 and a type 2; (c) a pattern of "whirling" golden triangles create triangles at smaller scales.

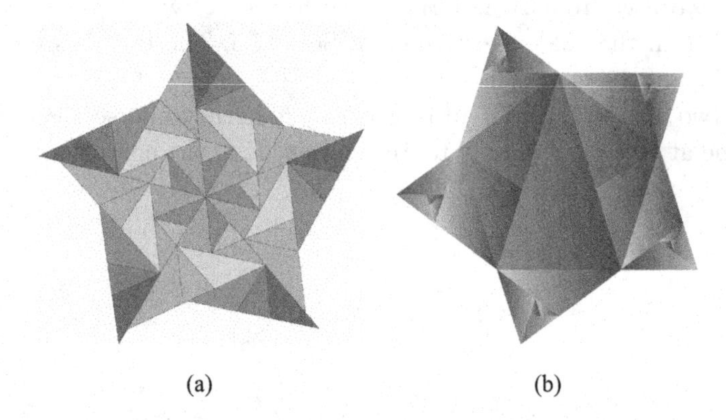

Fig. 17.9. Golden triangles designs by (a) Cindy Caldas; (b) Joseph Titterton.

17.8. Application of the Golden Section to Art and Architecture

The golden mean is related to human scale since your belly button divides your height approximately in the *golden section*, i.e., the ratio of 1: ϕ as shown in Fig. 17.10 for "Modulor Man." This is the symbol that the architect, LeCorbusier used to represent his proportional system based on the golden mean known as the Modulor [Kap1]. A construction of the golden section was described in Sec. 17.2 and illustrated in Fig. 17.2.

Exploration: The golden mean and art

The golden section has appeared in countless works of art as a natural center of tension. It often defines the central point of a painting which is slightly off from the geometric center so as to create a sense of tension or drama in the work of art. In some instances, the artist has intentionally introduced the golden section. In other cases, the golden section makes its appearance through some unconscious process on the part of the artist; it may be naturally programmed into our brains.

I invite you to explore great paintings to see if the central point of the painting divides the canvas into the golden section. To help you in this analysis, you can use the golden section line chopper in Fig. 17.11. You may reprint this figure and enlarge it.

Fig. 17.10. Trademark of LeCorbusier's Modulor illustrating the golden section.

The horizontal line on the top of the line chopper is divided in the golden section. From a print of a classic painting, take either its length or width and use the line chopper to find its golden section, and see if the central point of interest within the painting divides the canvas in the golden section.

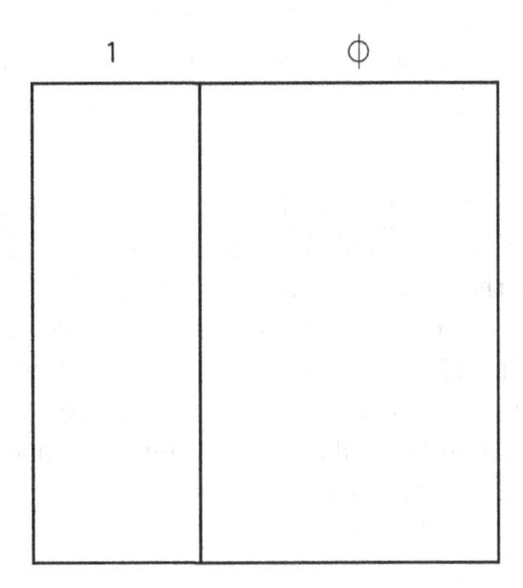

Fig. 17.11. A golden section line chopper.

17.9. A Regular Decagon and Regular Pentagon Within a Circle

Construction 6: Construct a regular decagon and regular pentagon within a circle

You are now able to inscribe a regular *decagon* inside a circle as follows:

a. Construct golden triangle 1 with base 1 unit and sides ϕ units (see Sec. 17.8). The vertex angle of this triangle is 36 deg., the central angle of the decagon. Therefore, the sides of Triangle 1 are radii of the circle and the base is a chord of the circle as shown in Fig. 17.12.

b. Mark off ten equal chord lengths to construct the sides and vertices of the decagon inscribed in the circle.

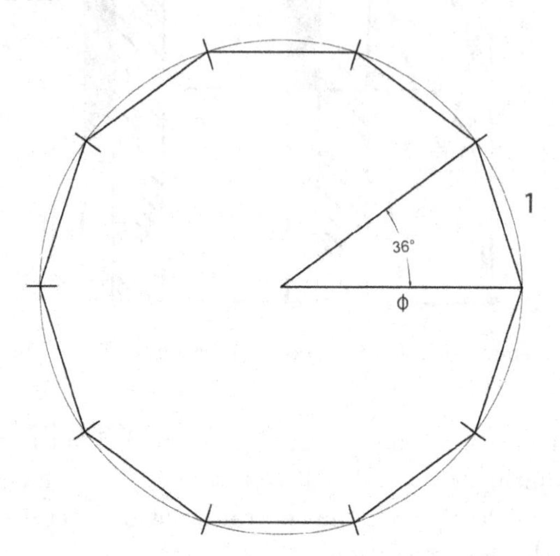

Fig. 17.12. Decagon inscribed in a circle.

A pentagon inscribed in a circle is constructed by connecting every other vertex.

17.10. Sacred Cut

In Fig. 17.13a, the compass point is placed at a vertex of a square and an arc is drawn through the center of the square. This arc divides the side of the square in what is known as the *sacred cut*, the ratio, $\theta : 1$ where $\theta = 1 + \sqrt{2}$ the silver mean [Kap2]. If four such arcs are drawn, one from each vertex of the square, the vertices of an *octagon* is the result as shown in Fig. 17.13b. In Fig. 17.13c, four such sacred cuts tile the square with three species of rectangles: a square S (1:1), a square root rectangle SR ($\sqrt{2}$:1), and a Roman rectangle RR (θ:1). A floor mosaic found in one of the Garden Houses of Ostia [Wat], the port city of the Roman Empire, in the form of this subdivision of a square is shown in Fig. 17.13d.

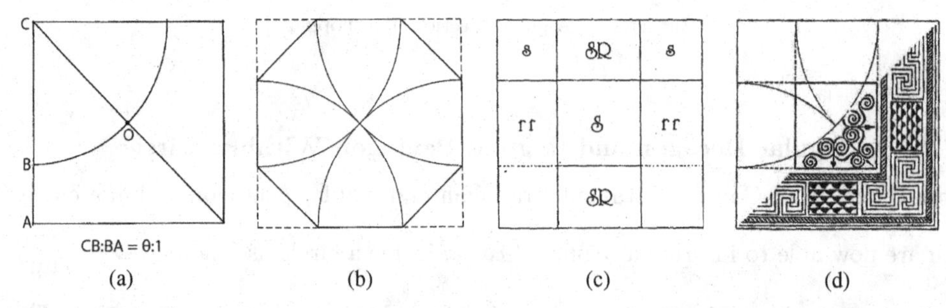

Fig. 17.13. (a) the sacred cut of an edge of a square; (b) four sacred cuts form the vertices of an octagons; (c) the sacred cut subdivides a square into a square (S), square root rectangle (SR), and a Roman rectangle (RR); (d) a floor mosaic found in one of the garden Houses of Ostia proportioned by the sacred cut.

These three rectangles have wonderful additive properties. Beginning with SR:

If you add a square, you get RR, i.e., SR + S = RR.
If you subtract a square, you also get an RR, i.e., *S*R − S = RR.
If you divide SR in half, you get two smaller SRs; i.e., SR = SR + SR.
If you add two SRs, you get a larger SR; SR + SR = SR.

These results are illustrated in Fig. 17.14 and follow from Fig. 16.7b, where two squares were removed from an RR resulting in an RR at a smaller scale.

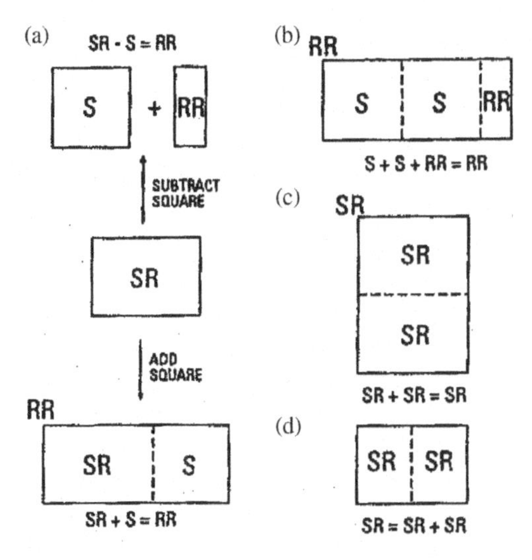

Fig. 17.14. The relationships between S, SR, and RR.

Construction 7: Construct an octagon from four sacred cuts of a square.

Construction 8: Cut out several S, SR and RR rectangles at three different scales and create a pleasing tiling with them. Two student examples are shown in Fig. 17.15.

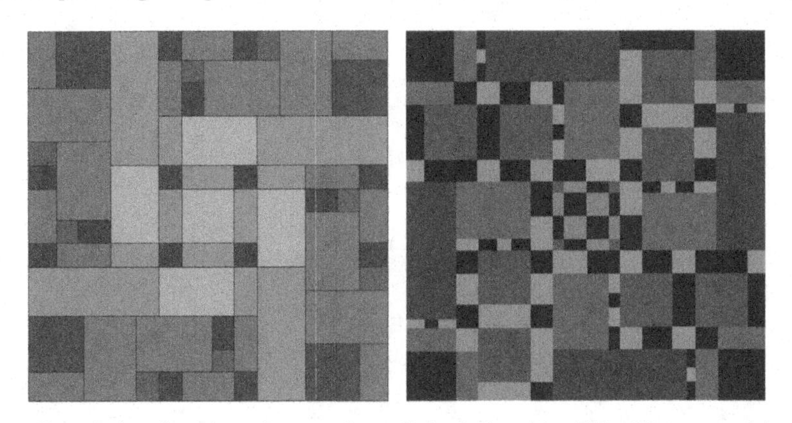

Fig. 17.15. Two designs based on the sacred cut (left: designed by Elliot Virtgaym; right: designed by Mark Bak).

CHAPTER 18

ISOMETRIES

18.1. Introduction

Geometry is about transformation. A geometrical figure is transformed. Its points change their positions, but certain aspects of the figure remain unchanged or *invariant*. The properties of geometry are determined by the nature of the transformations. The original figure and the transformed figure are considered to be equivalent, or in the case of Euclidean geometry, congruent. In this chapter, we will study in detail the transformations associated with Euclidean geometry called isometries.

Note: In this chapter and the next, we suspend our convention of using upper case letters to represent points and instead use lower case letters in order to enhance the clarity of the presentation.

18.2. Isometries

Isometries are transformations, *Tr*, that keep the distance between points *invariant*. As a result, they also preserve lengths and angles. We consider isometries of the plane in this chapter. There are exactly four classes of isometries in the plane: Translations, T, Rotations, S, Reflections, R, and Glide reflections, G, where we will consistently use T, S, R and G when referring to these isometries.

By the defining property of isometries, if p and q are two points of the plane and p' and q' are their images under some isometry, then the distance between p and q, $d(p, q)$, and the distance between their images, $d(p', q')$, is unchanged, i.e., $d(p, q) = d(p', q')$. Furthermore, we are guaranteed that there is at least one isometry T, S, R or G that maps p, q to $p,' q'$.

18.3. Translations

A *Translation*, T, is defined by specifying a vector $\vec{v}$ (see Chap. 15) and transforming each point in the plane by this vector. For a translation, no point remains in its original position, i.e., there are no *fixed points*. Figure 18.1a illustrates the translation of a stick

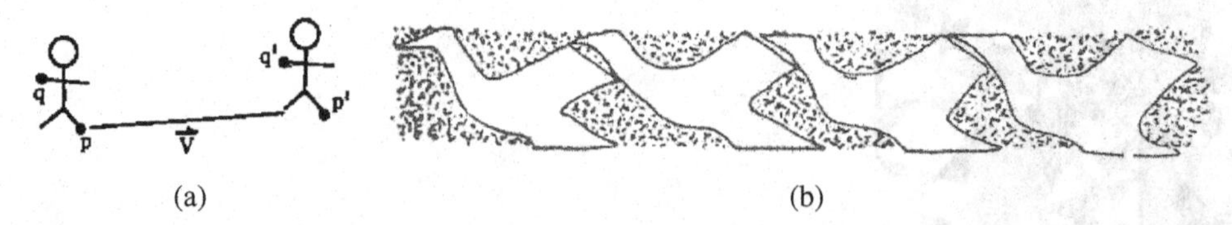

Fig. 18.1. (a) Translation of a figure in the plane; (b) Multiple translations of a Kala nut box, Nigeria.

figure, and Fig. 18.1b illustrates a design with multiple translations. In Fig. 18.1a, points p and q are transformed to p' and q'. Notice that the distance between p and q, $d(p,q)$ and the distance between p' and q', $d(p',q')$ are equal. (Why?) We will sometimes represent the vector characterizing the translation by a subscript to the translation symbol $T_{\vec{v}}$.

Example 1: In Sec. 15.10, we saw that any point p in the (x,y)-plane is represented by the vector $\vec{p}$ from the origin to the point. A translation in the plane can then be represented by adding vector $\vec{v} = (a,b)$, specifying the translation, to vector $\vec{p} = (x,y)$ specifying the point, so that,

$$T_{\vec{v}}\vec{p} = T_{(a,b)}(x,y) = (x+a, y+b) = \vec{p}' \tag{18.1}$$

e.g., if $(x,y) = (2,1)$ and $\vec{v} = (3,2)$ the transformed point is $\vec{p} = (5,3)$.

18.4. Rotations

A Rotation, S, is defined by specifying a *center of rotation* O and an angle θ. All points in the plane then rotate about this center by the given angle as shown in Fig. 18.2a. A design with multiple rotations is illustrated in Fig. 18.2b. In Fig. 18.2a, points p and q are transformed to p' and q'. Notice that $d(p,q) = d(p',q')$. (Why?) The center of rotation O is the only fixed point of the rotation. Sometimes we will denote the angle of rotation by a subscript and the center by a superscript to the rotation symbol S_θ^O.

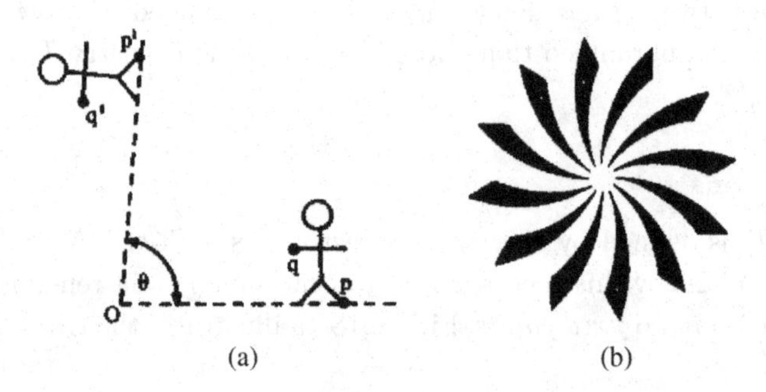

Fig. 18.2. (a) Rotation of a figure in the plane; (b) multiple rotations.

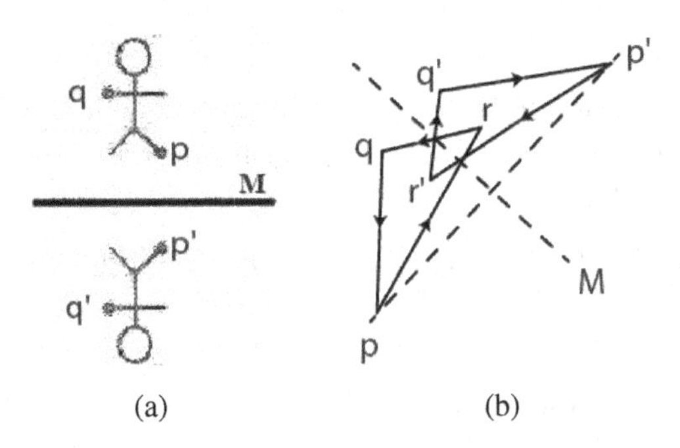

Fig. 18.3. Reflection of a figure in the plane.

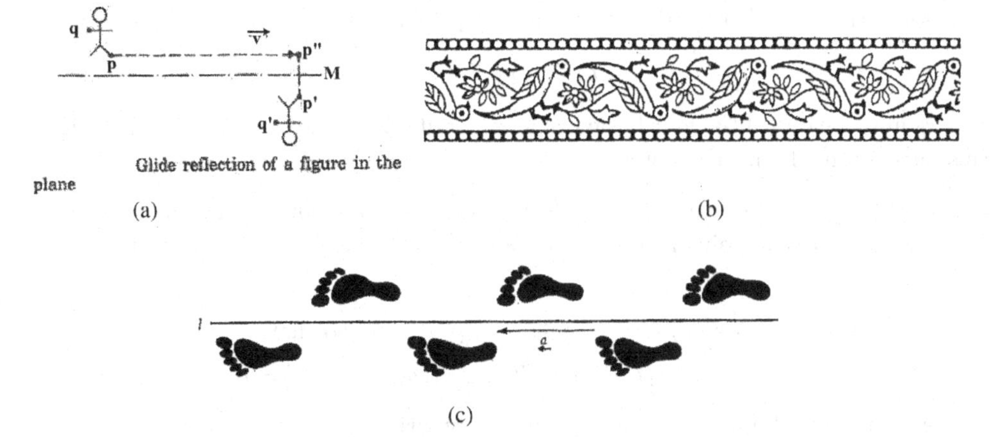

Fig. 18.4. (a) Glide reflection of a figure in the plane; (b) multiple glide reflections; (c) footprints along a line have glide reflection symmetry.

18.5. Reflections

A Reflection, R, is defined by specifying a *mirror-line*, M. Points on one side of M are projected perpendicular to the mirror to points equidistant on the other side of the mirror. This is illustrated in Fig. 18.3a. Figure 18.3b shows the reflection of a figure in the plane. In Fig. 18.3b, points p and q are transformed to p' and q'. Notice that $d(p,q) = d(p',q')$. (Why?) All points on the mirror are fixed.

18.6. Glide Reflections

A Glide reflection, G, is defined by specifying a mirror-line M and a translation vector $\vec{v}$ parallel to the mirror. A point in the plane is first translated by the vector and then reflected in the mirror-line as shown in Fig. 18.4a. A design with multiple glide reflections is shown in Fig. 18.4b. Another example of a glide reflection are footprints in the snow illustrated by Fig. 18.4c.

18:7. Identity Transformation

There is one additional isometry that should be mentioned. It is the transformation that leaves every point in space at the same position. This is known as the *identity* transformation, denoted by E.

Remark: The identity can be looked at as a translation by the zero vector or a rotation by the zero angle.

18.8. Compounding Isometries

We have been talking about the simple isometries T, S, R and G. However, since an isometry followed by another isometry must also be an isometry, (Why?) it is something of a detective story to discover the isometry that is the result of such a compound transformation. In other words, an isometry Tr_1, followed by Tr_2 yields Tr_3 written as,

$$Tr_2 \circ Tr_1 = Tr_3. \tag{18.2a}$$

So Tr_3 must be one of the four basic isometries, but which one is it? We will investigate this question in the next chapter.

Remark: The small circle in Eq. (18.2a) denotes composition and means that, when $Tr_2 \circ Tr_1$ operates on point P, the point transforms by the sequence of operations from right to left, i.e.:

$$Tr_1(p) = p'' \quad \text{and} \quad Tr_2(p'') = p' \quad \text{so that}$$
$$Tr_3 p = (Tr_2 \circ Tr_1)p = Tr_2(Tr_1(p)) = Tr_2 p'' = p'.$$

In the future, we eliminate the circle and simply write,

$$Tr_2 Tr_1 = Tr_3 \tag{18.2b}$$

and refer to this compound transformation as being the *product* of a pair of simple transformations. So we see here that transformations have a kind of algebra associated with them.

Example 2: Consider two translations, $T_{\vec{v}_1}$ and $T_{\vec{v}_2}$ corresponding to vectors $\vec{v}_1 = (2,1)$ and $\vec{v}_2 = (-1,3)$. Figure 18.5 shows point p being translated first by $T_{\vec{v}_1}$ to p'' and then by $T_{\vec{v}_2}$ to p', i.e., $(T_{\vec{v}_2} \circ T_{\vec{v}_1})p = p'$. We also see that the resulting transformation is $T_{v_1+v_2}$ where $\vec{v}_1 + \vec{v}_2 = (2,1) + (-1,3) = (1,4)$.

Example 3: A reflection followed by another reflection in the same mirror clearly leaves all points unchanged. (Why?) We state this as follows,

$$RR = E. \tag{18.3}$$

Example 4: The product of any transformation following or preceding the identity is just that transformation, i.e.,

$$Tr\,E = Tr = E\,Tr. \tag{18.4}$$

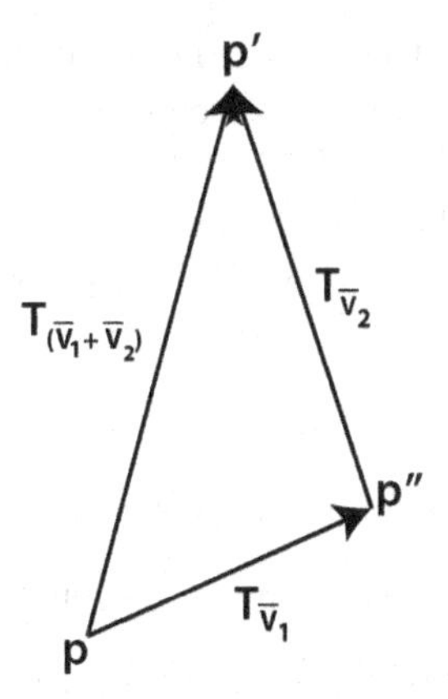

Fig. 18.5. (a), (b) The "product" of two translations is a translation.

18.9. Proper and Improper Isometries

The notion of a proper and improper isometry is important for determining which isometry is the result of compounding isometries. Consider the following fable:

In a universe called Flatlands lived a bunch of identical triangular shaped creatures as shown in Fig. 18.6a. Whenever two Flatlands creatures met, they engaged in the ritual of matching up their bodies point for point by movements in the plane that consisted of a series of translations and rotations. They were always successful in doing this. All was well until one day an alien creature appeared on the scene in Fig. 18.6b. No matter how much twisting and turning in the planar Flatlands universe, he could not match up with the others. However, a Flatlands Einstein arose and showed that he could indeed be matched up by removing him from the plane into the previously unknown three-dimensional space, inverting him, and placing him back into the plane. The transformation achieved by removing him from the plane and inverting, sometimes called a "flip," is equivalent to either a reflection or glide reflection.

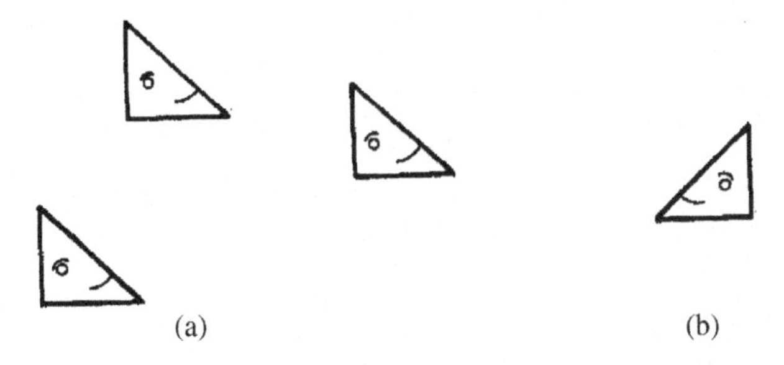

Fig. 18.6. Three Flatland creatures encountering an alien.

The movements available to the Flatland creatures, i.e., translations and rotations are called *proper isometries*, P. The flips, reflections and glide reflections, are called *improper isometries*, I.

It is easy to see that a P followed by another P results in a P, while an I followed by an I results in a P, and I followed by P or P followed by I is an I. Summarizing these relationships,

$$PP = P, \quad II = P, \quad IP = I, \quad PI = I. \tag{18.5}$$

Remark: Even and odd integers follow similar rules since:

$$\text{even} + \text{even} = \text{even}, \quad \text{odd} + \text{odd} = \text{even}, \quad \text{even} + \text{odd} = \text{odd} = \text{odd} + \text{even}.$$

As a result, proper transformations are sometimes referred to as even, and improper as odd.

Remark: An equivalent situation to the Flatland creatures exists in our own universe. A left glove cannot be made to match up with a right glove. However, if a modern day Einstein came along, he would be able to show that if the left glove could be lifted into the fourth dimension, it could then be flipped in 4-D and returned to 3-D as a right glove.

Isometries are uniquely determined by the transformations of three non-collinear points (not all points on the same line) as proven in Appendix 18A. So if you know how the vertices of a triangle transform, that completely determines the isometry. However, if the images of two points are given, exactly two isometries will do the job; one isometry will be proper and the other improper. If only one image is given, you can find an infinite number of isometries that will transform the given point to its image.

Problem 1: Two points and their images lie at the vertices of a parallelogram in Fig. 18.7a. $p = (0,0)$, $p' = (2,4)$, $q = (4,0)$, $q' = (6,4)$. Certainly, this transformation can be carried out by a translation. What is the translation vector? Your job is to show how it can also be carried out by a glide reflection. What is the equation of the mirror line of the glide reflection?

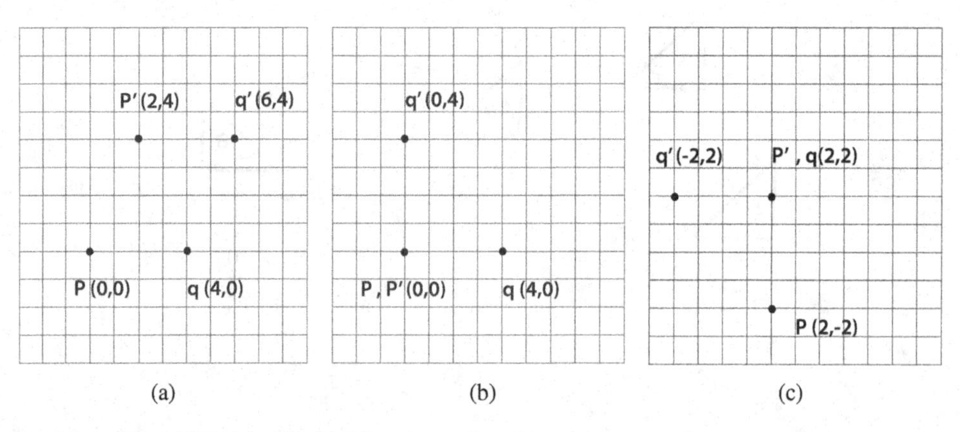

Fig. 18.7. (a), (b), (c) Three examples of isometries mapping p, q to p', q'.

Problem 2: In Fig. 18.7b, $p = p' = (0,0)$ while $q = (4,0)$ and $q' = (0,4)$. This can certainly be carried out by a 90-deg. rotation. What improper transformation will also carry out this isometry? What is the equation of the mirror line?

Problem 3: In Fig. 18.7c, $p = (2,-2)$, $p' = (2,2)$, $q = (2,2)$, and $q' = (-2,2)$. Show that this isometry can be carried out by a rotation. Where is the center? What is the angle? Then show that it can also be carried out by a glide reflection. Where is the mirror-line? What is the translation vector?

18.10. Additional Problems

1. Suppose T is an unknown translation, but you know that $T(p) = q$ for known points p and q. How many possibilities are there for T?
2. Suppose T is an unknown translation for which you do not know the exact location of $T(p)$ but you do know the direction of the translation (although not its length) and that $T(p)$ lies on a given line. How many possibilities are there for T?
3. Suppose S is a rotation about some unknown center by some unknown angle. Suppose there are known points p and q where $q = S(p)$.

   (a) Explain why you cannot determine where the center is.
   (b) There is a construction you can carry out that narrows down where the center might be. What is it?
   (c) How can you find the actual center?

4. Suppose R is an unknown reflection, but you know that $R(p) = q$ for known points p and q. How many possibilities are there for R?
5. (a) Consider a rectangle that is not a square. List as many isometries as you can that keep the rectangle unchanged although the points within it can change their positions.
   (b) Answer this question if the rectangle is a square.
6. Answer "true" or "false" for the following statements. Explain your answer.

   (a) "Circlehood" is an invariant of any isometry.
   (b) The slope of a line is an invariant of any isometry.
   (c) If $\Delta\ ABC$ transforms to $\Delta\ A'B'C'$, by an isometry and D is the center of the circle inscribed in triangle ABC, then D' is the center of the circle inscribed in triangle $A'B'C'$.

7. Give an example of an isometry that transforms the x-axis into the y-axis and has the origin as a fixed point — that is, the origin corresponds to itself under the isometry. Is there just one such isometry or can you find another?
8. Give an example of an isometry that preserves verticality, that is, the image of each vertical line is a vertical line. Can you find a second example? A third?
9. Is verticality of lines an invariant of all isometries?
10. Suppose a certain isometry transforms the y-axis to itself (but individual points on the axis are not necessarily invariant). What can you say about the image of the x-axis?

11. If, for a transformation, Tr, parallel lines are invariant, does it follow that Tr is an isometry?

Appendix 18A. Proof that an Isometry is Determined by the Transformation of Three Points [Mey]

Lemma 18A.1: *Three circles with non-collinear centers can only intersect at a single point as illustrated by point A in Fig.* 18A.1a. *If the centers are collinear, the circles can intersect at two points as shown by points A and B in Fig.* 18A.1b.

Remark: A lemma is a theorem that plays a supporting role in the proof of another theorem.

Theorem 18A.1: *An isometry is determined by the transformation of three non-collinear points.*

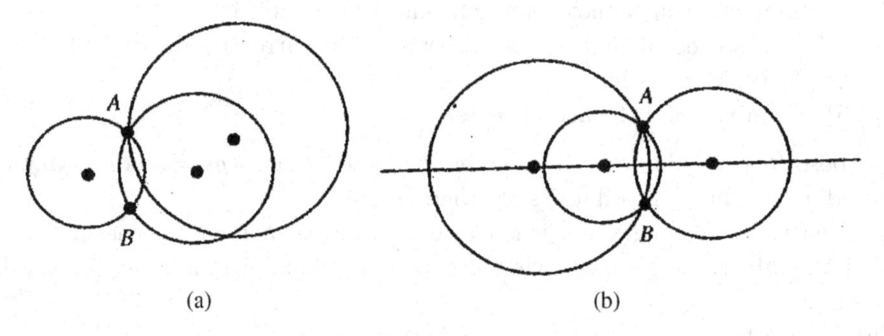

Fig. 18A.1. Three circles that intersect do so in two points only if the centers are collinear.

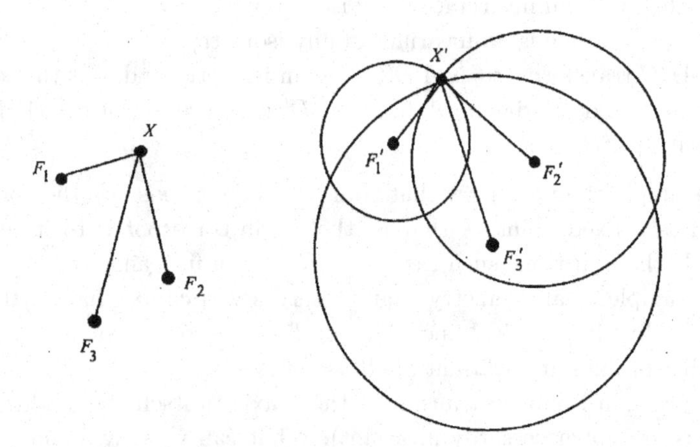

Fig. 18A.2. An isometry is uniquely determined by its effect on three non-collinear points.

Proof. The proof is by contradiction.

1. Assume that there are two isometries, Tr_1 and Tr_2 such that $Tr_1 \neq Tr_2$, and Tr_1 and Tr_2 map $F_1, F_2, F_3 \rightarrow F_1', F_2', F_3'$ for non-collinear F_1, F_2, F_3 (see Figs. 18A.2a, and 18A.2b).
2. F_1', F_2', F_3' are non-collinear since F_1, F_2, F_3 forms an angle and angles are preserved by isometries.
3. There exists an x such that $Tr_1 x = x'$ and $Tr_2 x = x''$ for $x' \neq x''$ since $Tr_1 \neq Tr_2$.
4. Let $d(F_1, x) = r_1$, $d(F_2, x) = r_2$, $d(F_3, x) = r_3$ (see Fig. 18A.2a).
5. $d(F_1', x') = d(F_1', x'') = r_1$, $d(F_2', x') = d(F_2', x'') = r_2$, $d(F_3', x') = d(F_3', x'') = r_3$ since isometries preserve distance.
6. Points x' and x'' each lie on three circles with non-collinear centers at F_1', F_2', F_3' and radii r_1, r_2, r_3 yet $x' \neq x''$.
7. By Lemma 18A.1, there can only be one point where three circles intersect so we have reached a contradiction.
8. Therefore $Tr_1 = Tr_2$, and the transformation is unique and completely determined by the mapping of three non-collinear points.

QED

Remark: Consider the mapping of two points by an isometry, i.e., $F_1, F_2 \rightarrow F_1', F_2'$. By an argument similar to the one above and according to Lemma 18A.1, there can be two isometries given that only two points are mapped. As we saw in Sec. 18.9, one of these isometries is proper and the other is improper.

Remark: Given three non-collinear points A, B, C we can create a triangle with the three given sides, AB, BC, CA. Therefore, any isometry applied to A, B, C results in a congruent triangle A', B', C' by SSS. By Theorem 18A.1, we now see that the transformation of any figure is completely determined and congruent under isometries. For this reason, the subject of geometry can be studied by focusing primarily on triangles.

CHAPTER 19

ISOMETRIES AND MIRRORS

19.1. Introduction

Isometries and mirrors are closely connected. In fact, all isometries can be carried out by simultaneous reflections in no more than three mirrors. The proof of this is given in Appendix 19A. In this chapter, we show how isometries and their products can be carried out with mirrors.

19.2. Reflections in One Mirror

Given one mirror, the isometry specified by this mirror is a simple reflection as shown in Fig. 19.1.

19.3. Reflections in Two Mirrors

If two mirrors, M_1 and M_2, intersect at point O at angle θ; the isometry defined by them is a rotation about O through angle 2θ as shown in Fig. 19.2a for an arbitrary point p and its image, p', i.e.,

$$(R_2 R_1)p = S_{2\theta}^{O} p = p' \tag{19.1}$$

using the notation of Eq. (18.2). This is consistent with Sec. 18.9 since reflections are improper and the product of two improper transformations is proper, and since there is a fixed point at O, the result must be a rotation.

If the two mirrors are rotated about O to a new position maintaining the same angle of intersection, θ, the location of the image point is unchanged as shown in

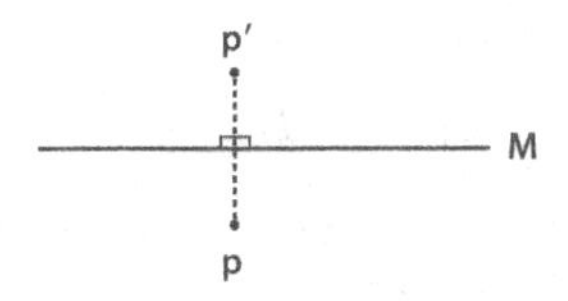

Fig. 19.1. Reflection in 1 mirror.

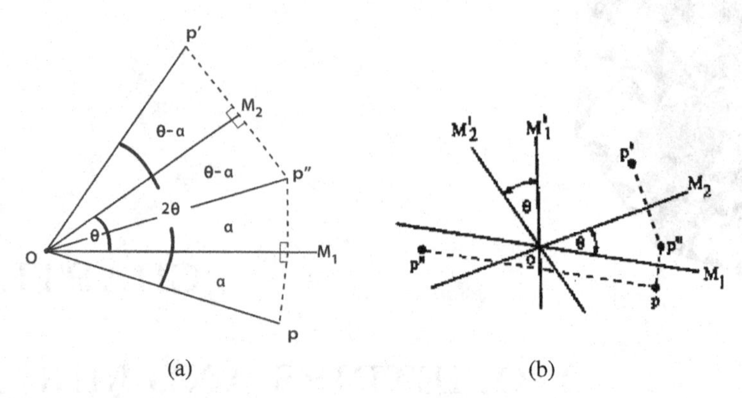

Fig. 19.2. (a) Multiple reflections in two mirrors intersecting at a point; (b) the multiple reflection is unaffected by the position of the mirror pairs.

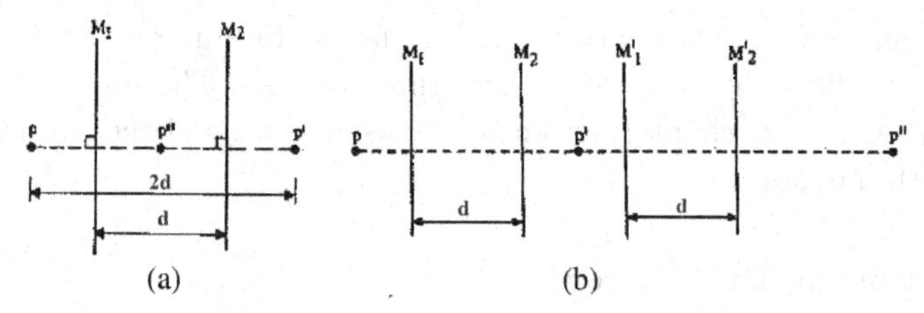

Fig. 19.3. (a) Multiple reflections in two parallel mirrors; (b) the multiple reflection is unaffected by the position of the mirror pairs.

Fig. 19.2b. If the reflections are taken in reverse order, the result is a rotation through 2θ with opposite orientation, i.e.,

$$R_1R_2 = S^O_{-2\theta}.$$

Positive values of θ correspond to counterclockwise rotations, while negative values are clockwise.

If the mirrors are parallel, then the isometry is a translation perpendicular to the mirror-lines through twice the distance, d between the mirrors as shown in Fig. 19.3, i.e.,

$$(R_2R_1)p = T_{(2d,0)}p = p' \tag{19.2}$$

using the vector notation of Chap. 15. Again, if the mirrors are moved to a new location in a parallel direction and still d units apart, the image point is unchanged. Again, if the reflections are carried out in the reverse order, the direction of the translation is reversed, i.e.,

$$R_1R_2 = T_{(-2d,0)}.$$

19.4. Reflections in Three Mirrors

There are five distinct ways in which three mirror lines can be oriented, and they are shown in Fig. 19.4.

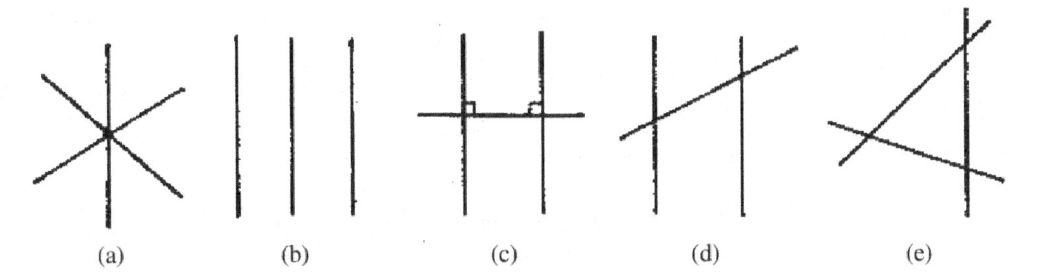

Fig. 19.4. Five ways in which three mirrors can intersect.

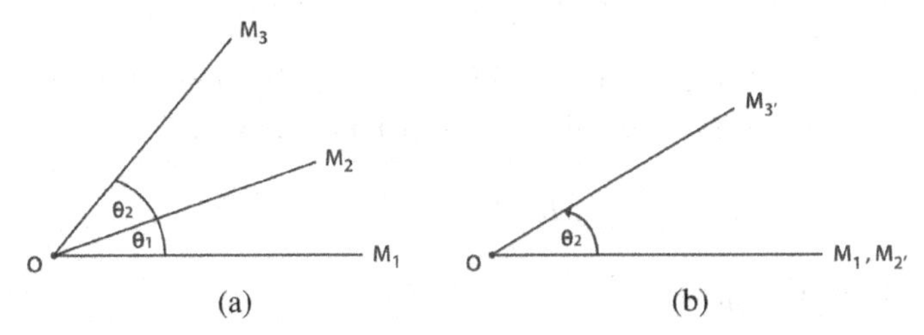

Fig. 19.5. (a) Multiple reflections in three mirrors meeting at a point; (b) the resultant reflection is determined by moving M_2 and M_3 so that M_2 coincides with M_1.

19.4.1 Three mirrors intersecting at a common point

The multiple reflection in mirrors M_1, M_2, M_3 intersecting at point O, shown in Fig. 19.4a, and with more detail in Fig. 19.5a, can be expressed as:

$$R_3(R_2R_1) \quad \text{or} \quad (R_3R_2)R_1. \tag{19.3a}$$

Here M_2 is oriented at angle θ_1 with respect to M_1, and M_3 is oriented at θ_2 with respect to M_2.

Without the proof, we state that the associative law holds for the product of isometries so that Expressions (19.3a) have the same result, and in the future we will simply state its product as,

$$R = R_3R_2R_1. \tag{19.3b}$$

Such a transformation is the product of three improper transformations, and so by Eqs. (18.5), the result must be improper, and since O is a fixed point of this transformation, the improper transformation must be a reflection (since glide reflections have no fixed points), and the mirror-line of this reflection must go through O. To find the angle at which the mirror-line is oriented to mirror 1, rewrite Eq. (19.3) as,

$$(R_3R_2)R_1. \tag{19.4}$$

Since by Eq. (19.1), the product R_3R_2 can be expressed equally well by rotating mirrors M_2 and M_3 clockwise to $M_{2'}$ and $M_{3'}$, maintaining angle θ_2 between them, until $M_{2'}$ coincides with M_1. Then $M_{3'}$ in its new position is at an angle of θ_2 with respect to

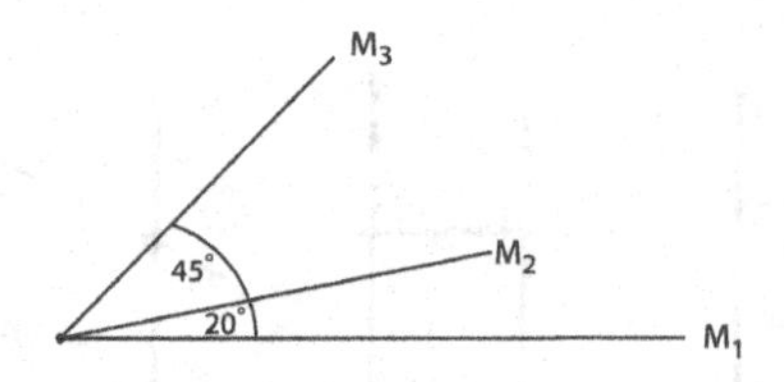

Fig. 19.6. Where is the resulting mirror for this transformation?

M_1 as shown in Fig. 19.5b. Now Eq. (19.4) can be rewritten,

$$R = R_{3'}(R_{2'}R_1) = R_{3'}(R_1R_1) \tag{19.5}$$

since $R_{2'} = R_1$. But a reflection followed by another reflection in the same mirror has no effect on a given point; it is the identity transformation E, i.e.,

$$R_1R_1 = E. \tag{19.6}$$

Replacing Eq. (19.6) in Eq. (19.5),

$$R = R_{3'}E = R_{3'}$$

where the identity transformation has been eliminated since it has no effect. So we see that this multiple reflection is equivalent to a reflection in mirror $M = M_{3'}$ through point O at angle θ_2 with respect to mirror M_1.

Problem 1: For the three mirrors shown in Fig. 19.6, determine the location of the mirror-line of the reflection that is the result of this multiple reflection.

19.4.2 Three parallel mirrors

Three parallel mirrors are shown in Fig. 19.4b. Since parallel lines can be imagined to intersect at a point at infinity, the analysis of this case is similar to Sec. 19.4.1. We illustrate this situation with the example shown in Fig. 19.7a. The three mirrors are placed in an (x, y)-coordinate system oriented perpendicular to the x-axis. M_1 is placed at $x = 3$, M_2 at $x = 8$, and M_3 at $x = 12$ so that the distance between M_1 and M_2 is 5 units and the distance between M_2 and M_3 is 4 units. We wish to find the

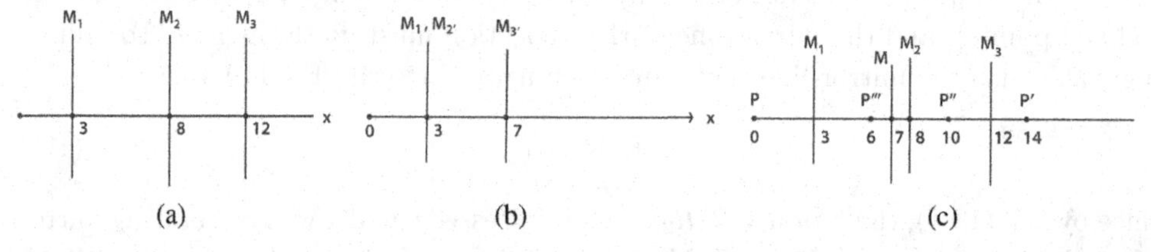

Fig. 19.7. (a) Multiple reflections in three parallel mirrors; (b) M_2 and M_3 are moved so that M_2 coincides with M_1; (c) the multiple reflection is shown by following the trajectory of a point.

isometry R that is the result of,

$$R = R_3 R_2 R_1.$$

Since we have again the product of three improper transformations, the result will be improper. But since the mirrors can be thought of as intersecting at a point at infinity, that point is fixed so that the resulting transformation must be a reflection since glide reflections have no fixed points. We proceed to find the mirror line M of that reflection.

As we did for three mirrors intersecting at a point, we move M_2 and M_3 to $M_{2'}$ and $M_{3'}$, preserving the distance between them, so that $M_{2'}$ coincides with M_1 as shown in Fig. 19.7b. As before, this has no effect on the transformation so that,

$$(R_3 R_2) R_1 = R_{3'}(R_{2'} R_1) = R_{3'}(R_1 R_1).$$

Again, since

$$R_1 R_1 = E,$$

it follows that

$$R = R_{3'},$$

and the resulting mirror-line is $M = M_{3'}$ located at $x = 7$ units (4 units apart from M_1).

This problem can be solved in another way. Begin with a typical point p located at $x = 0$ as shown in Fig. 19.7c. Reflect p first in M_1 to $x = 6$, then in M_2 to $x = 10$ and finally in to $x = 14$. In other words, the original point p is at $x = 0$ and its image p' is at $x = 14$. Since we know that this is carried out by a reflection, the mirror M must be located at $x = 7$ as we previously discovered.

Problem 2: Find the location of the mirror-line for reflections in three parallel mirrors placed at $x = 2$, $x = 8$ and $x = 10$. Analyze this problem by both methods Sec. 19.4.2.

19.4.3 Reflections in three mirrors resulting in a glide reflection

Figure 19.4c clearly results in a glide reflection. The analyses of Figs. 19.4d and 19.4e are more complex. They also represent glide reflections as shown in Appendix 19B.

19.5. Multiple Rotations

In this section, we consider point A to be located in a Cartesian coordinate system at $(15, 0)$ while point B is at the $(0, 0)$ and point p is at $(10, -5)$.

Consider two rotations S^A_{70} and S^B_{-30}. S^A_{70} is a rotation about point A by an angle of 70 deg. counterclockwise while S^B_{-30} is a rotation about B by an angle of 30 deg. clockwise. (Remember: Positive angles are counterclockwise and negative angles are clockwise.)

Begin with point p and first perform S^A_{70} resulting in p''. Then follow this with S^B_{-30} resulting in p' as shown in Fig. 19.8. The product of these rotations is either a rotation or a translation. (Why?) In this case it is a rotation S^C_θ about unknown point

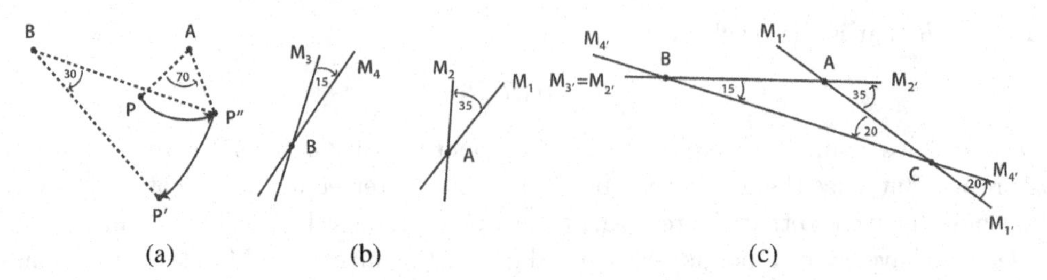

Fig. 19.8. (a) Point p transforms to p' by rotations about two different points A and B; (b) the multiple rotation is carried out with mirrors; (c) the fixed point C and the angle θ of the resulting rotation M_θ^C is determined by moving the mirrors so that mirror $M_{2'}$ and $M_{3'}$ coincide with the line between A and B.

C through unknown angle θ, i.e.,

$$S_{-30}^B S_{70}^A = S_\theta^C. \tag{19.7}$$

(Remember: The first operation is always placed on the right of this "product" and the second operation is to the left.) But where is the point of rotation C, and what is the angle θ of rotation? The answer follows from geometry using what I call the *Method of Multiple Mirrors.* Represent S_{70}^A by the pair of mirrors M_1 and M_2 as shown in Fig. 19.8b inclined at an angle 35 deg. (Why?), i.e.,

$$S_{70}^A = R_2 R_1. \tag{19.8}$$

Notice that mirror M_2 is oriented counterclockwise 35 deg. with respect to M_1 since we wish to generate a counterclockwise rotation of 70 deg.

Now carry out S_{-30}^B with the pair of mirrors, M_3 and M_4 inclined at an angle 15 deg. as shown in Fig. 19.8b to give:

$$S_{-30}^B = R_4 R_3. \tag{19.9}$$

Here M_4 is oriented 15 deg. clockwise with respect to M_3 since we wish to achieve a 30 deg. clockwise rotation.

Replacing Eqs. (19.8) and (19.9) in Eq. (19.7), the pair of rotations are represented by,

$$S_\theta^C = R_4 R_3 R_2 R_1.$$

Since any pair of mirrors inclined at a given angle has the same result on point P, we rotate M_1 and M_2 to $M_{1'}$ and $M_{2'}$; and M_3 and M_4 to $M_{3'}$ and $M_{4'}$ so that $M_{2'}$ and $M_{3'}$ coincide with the line joining A and B, i.e. $M_3' = M_2'$. The rotations S_{70}^A and S_{-30}^B are then generated as follows,

$$S_{70}^A = R_{2'} R_{1'} \quad \text{and} \quad S_{-30}^B = R_{4'} R_{3'} = R_{4'} R_{2'},$$

where mirror-lines $1'$ and $4'$ are shown in Fig. 19.8c. As a result, we now have that,

$$\begin{aligned} S_\theta^C &= S_{-30}^B S_{70}^A \\ &= R_{4'} R_{2'} R_{2'} R_{1'} \end{aligned}$$

Since a reflection followed by another reflection in the same mirror is equivalent to the identity, it follows (see Eq. (18.3)) that $R_{2'}$ followed by $R_{2'}$ is the identity transformation,

$$R_{2'}R_{2'} = E.$$

and therefore,

$$S_\theta^C = R_{4'}R_{1'}.$$

Figure 19.8c shows that C lies at the intersection of $M_{1'}$ and $M_{4'}$. Notice that the external angle of ΔABC is 35 deg. while the alternate interior angles of ΔABC are 15 deg. and ϕ. Since the external angle of a triangle equals the sum of the alternate interior angles (see Theorem 8.1),

$$15 + \phi = 35 \quad \text{or} \quad \phi = 20\,\text{deg}.$$

Since the angle of rotation is $\theta = 2\phi$ (Why?),

$$\theta = 40\,\text{deg}.$$

Problem 3: For the multiple rotation, $S_{-40}^B S_{80}^A$:

a. Use compass and straightedge to find the image p' of point p given points A and B also shown in Fig. 19.9.
b. Since

$$S_{-40}^B S_{80}^A = S_\theta^C,$$

use the Method of Multiple Mirrors to find C and θ.
c. Does the result of parts (a) and (b) depend on the order in which the rotations are carried out, i.e., does,

$$S_{-40}^B S_{80}^A = S_{80}^A S_{-40}^B?$$

If not, what are the new values of C and θ?

Problem 4: Repeat Problem 3 for the product of rotations given by, $S_{90}^B S_{90}^A$.

Problem 5: Carry out the product of rotations given by, $S_{180}^B S_{180}^A$.

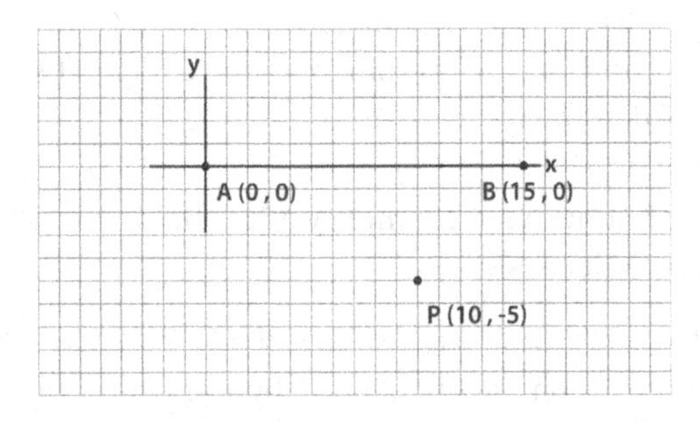

Fig. 19.9. The grid to solve Problem 3.

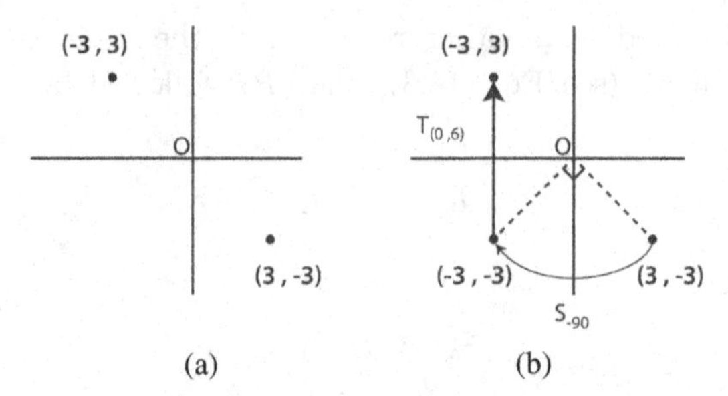

Fig. 19.10. Transform $p(3,-3)$ to $p'(-3,3)$ by a combination of a 90 deg. clockwise rotation about O followed by a translation upwards by 6 units. The result is a rotation of angle θ about C. Where is C? and what is the angle θ?

Show that the product is a translation. It turns out that when the angles of the rotation sum to 360 deg., the product results in a translation. What is the vector of the translation?

Challenge. Problem 6:

In Fig. 19.10, a point at $(3,-3)$ is transformed to $(-3,3)$.

There are an infinite number of isometries that carry out this transformation. We state three of them:

a. Rotation of 180 deg. about the origin O as shown in Fig. 19.10a,

$$S_{180}(3,-3) = (3,-3).$$

b. A translation with vector $\bar{v} = (-6,6)$,

$$T_{(-6,6)}(3,-3) = (-3,3).$$

c. A clockwise rotation of 90 deg. about the origin to $(-3,-3)$ followed by a translation through six units upwards, i.e, $\vec{v} = (0,6)$, as shown in Fig. 19.11,

$$T_{(0,6)}S_{-90}(-3,-3) = (-3,3).$$

Since the product of a translation and a rotation must be a proper transformation, the result is either a rotation or a translation. Show that this product is a rotation. Where is the center C located, and what is the angle?

Hint: Represent the rotation by a pair of mirrors, M_1 and M_2 and the translation by a pair of mirrors M_3 and M_4. If the mirrors are reoriented so that $M_{2'}$ and $M_{3''}$ coincide, the rotation is determined by $M_{1'}$ and $M_{4'}$.

Appendix 19A. Proof that Any Isometry Can be Carried Out with No More than Three Mirrors

Theorem 19A.1: *Any isometry can be carried out with no more than three mirrors.*

Proof. Since an isometry is completely defined by the transformation of three points we need only show that given two congruent triangles, ΔABC and $\Delta A'B'C'$, shown in

Fig. 19A.1 (shaded), ΔABC can be transformed to $\Delta A'B'C'$ by no more than three mirror reflections.

1. Let mirror M_1 be placed so as to reflect vertex A to vertex A' and ΔABC to $\Delta A'B''C''$. M_1 is the perpendicular bisector of AA'.
2. Place mirror M_2 so as to reflect vertex B'' to B' and $\Delta A'B''C''$ to $\Delta A'B'C'''$. M_2 is the angle bisector of $< B'A'B''$.
3. Finally place mirror M_3 on $A'B'$ to reflect C''' to C' and triangle $\Delta A'B'C'''$ to $\Delta A'B'C'$.

So three mirrors have been used to reflect ΔABC to $\Delta A'B'C'$ and we are done.

QED

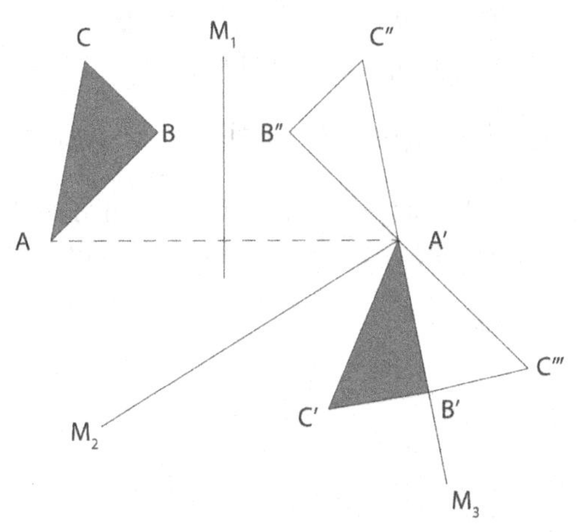

Fig. 19A.1. Any isometry can be carried out using three mirror reflections. Beginning with two congruent triangles, vertices A, B, C are transformed to A', B' and C'.

Appendix 19B. Three Mirrors in a Configuration of that Results in a Glide Reflection

Note: Points will be indicated by lower case letters.

Three mirrors (see Fig. 19B.1a) show how point p is reflected in mirrors M_1, M_2 and M_3 to point p', i.e.,

$$p' = (R_3R_2)R_1p.$$

Looking at Fig. 19B.1a, you see p first reflected in M_1 to p''. Since M_2 and M_3 are inclined at point O by some angle θ, p'' rotates clockwise about O by angle 2θ to p'. (Why?)

In Fig. 19B.1b, mirrors M_2 and M_3 are rotated about point O until M_2' is perpendicular to M_1 (Why can this be done?) and M_3 is rotated to M_3'.

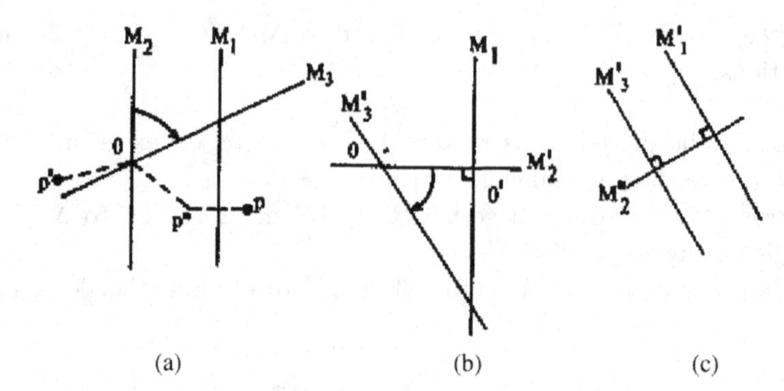

Fig. 19B.1. (a), (b), (c) Multiple reflections in three mirrors produces a glide reflection.

Reflections satisfy the associative law of mathematics so that,

$$(R_3R_2)R_1 = R_3(R_2R_1).$$

Now rotate M_1 and M_2' about point O' until M_2' is perpendicular to M_3' as in Fig. 19B.1c.

We see that mirrors M_1, M_2, M_3 have been transformed to M_1', M_2', M_3', the standard positions of a glide reflection, i.e., a translation in mirrors M_1', M_3' and a reflection in M_2'. QED

CHAPTER 20

KALEIDOSCOPES AND SYMMETRY

20.1. Introduction

When you think of symmetry, you think of mirrors. Figure 20.1 shows exact bilateral symmetry in a Senufo wooden mask from the Ivory Coast. One half of the face reflects exactly to the other half in a mirror.

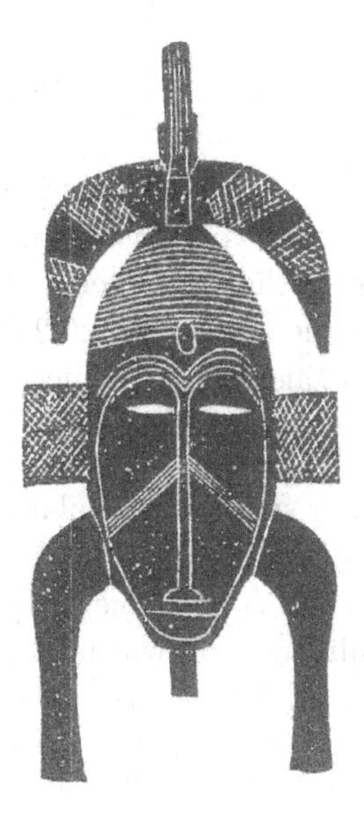

Fig. 20.1. Bilateral symmetry in a Senufo wooden mask, Ivory Coast/Mali/Upper volta.

Do the following exercises. You will need scissors, blank paper, marker, two small rectangular mirrors, some sequins or other small attractive objects and, a picture of your face facing forward.

20.2. Exercises

Exercise 1: Your face appears to be symmetric. Let us see how symmetric it is. Place a mirror along the line of symmetry that divides the left side of a photograph of your

Fig. 20.2. A pattern with kaleidoscope symmetry.

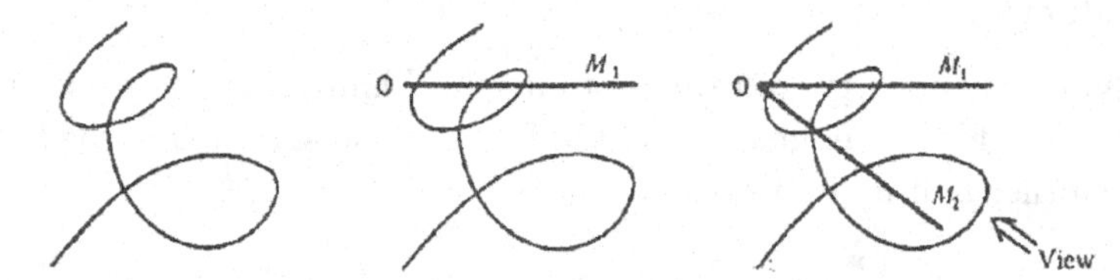

Fig. 20.3. Place this smooth curve between two mirrors and move one mirror until the curve repeats.

face from the right side and see whether half of your face and its mirror image combine to give a realistic or distorted image of your actual face.

Symmetry around a point is familiar in patterns observed in the central spaces of buildings and in patterns such as the one in Fig. 20.2. It is also the kind of symmetry that you see when you look into a kaleidoscope, which is why these patterns are called kaleidoscope symmetries.

Exercise 2: Place one corner of a small rectangular mirror at point O in Fig. 20.3, and place the edge of the mirror on line M_1. Now place the corner of a second mirror at O and vary its angle with respect to M_1 until the image of the curve between the two mirrors exactly repeats in the mirrors. You will notice that this occurs at a discrete set of angles. What are they? Record four of the angles below.

Angle 1 = ______ deg., Angle 2 = ______ deg.,

Angle 3 = ______ deg., Angle 4 = ______ deg.

Do you see a pattern to these angles?

Exercise 3: To discover how a kaleidoscope works, place the two small rectangular mirrors perpendicular to the paper protractor in Fig. 20.4, and open them to a sequence of angles, $\theta = 180/n$ deg. for $n = 1, 2, 3, 4, 5, 6, 8$. Place one colored sequin or just a pencil point mark between the mirrors and note the number of sequin images (including the original) that appear in the mirrors as a function of the angle of intersection. Record this information in Table 20.1.

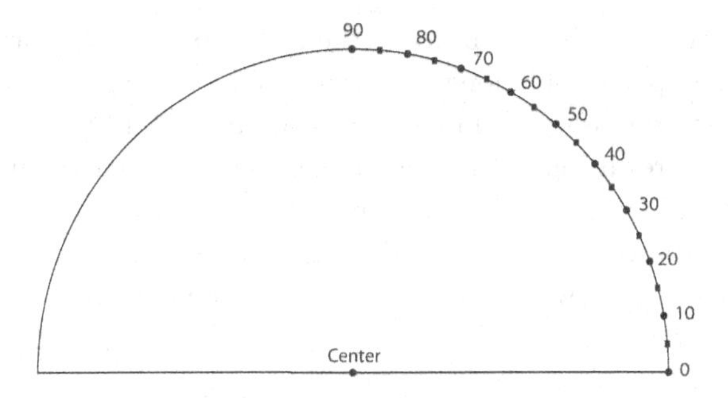

Fig. 20.4. A protractor to measure the angle between two mirrors.

Table 20.1.

n	$\theta = \frac{180}{n}$ deg.	Number of sequins
1	180	
2	90	
3	60	
4	45	
5	36	
6	30	
8	22.5	

Fig. 20.5. Snowflake pattern.

Now place several sequins or other attractive objects between the two mirrors open to one of the angles in Table 20.1 and notice the lovely patterns that appear between the mirrors.

All actual snowflakes exhibit hexagonal symmetry as illustrated by the pattern shown in Fig. 20.5. In the next exercise, I invite you to create your own "snowflake" pattern. (It need not be hexagonal.)

Exercise 4: Fold a piece of construction paper in half. Draw a line oblique to the fold of the paper at an angle of $180/n$ (choose $n = 4$ or 45 deg. and one other of these angles). Fold along this line. Next, cut away the excess paper (the portion of the paper that does not double up). Refold along the original fold, and again cut away the excess paper. Continue this process of folding and cutting until there is no longer any excess paper. Next, fold the resulting figure into a multilayered triangle and cut a pattern or motif into the triangle. Unfold to get a snowflake pattern around the central point consisting of several rotations and reflected images of your motif that are similar to an actual snowflake. Can you devise a way of creating snowflake patterns with rotations but no reflections of your motif?

Now that you have created a snowflake pattern for $n = 4$ or 45 deg., fold the pattern into the 45 deg. sector and place it between two mirrors at 45 deg. Do you notice that you obtain the entire unfolded snowflake pattern?

CHAPTER 21

GROUPS AND KALEIDOSCOPE SYMMETRY

21.1. Groups and Symmetry

We have seen in Chap. 18 that isometries are transformations of points in space that preserve the distance between points and angles between lines. Euclidean geometry can be thought of as the study of geometric figures that are invariant under isometries. On the other hand, in projective geometry, figures are transformed by projecting them through a point or a series of points. Such transformations do not preserve length and angle but do preserve a quantity called cross-ratio (see Appendix A1). This notion of a set of transformations preserving some property of space pervades all of mathematics.

The idea of *symmetry* is also bound up with the notion of preservation of a pattern under a set of transformations, and for Euclidian geometry these symmetries are invariant under isometries. Points of the pattern transform to new points. However, the pattern remains unchanged. For example, consider the letter 'A'. In Fig. 21.1a, notice that 'A' is preserved under a reflection in a vertical line, R_V. Of course, it is also preserved by the identity transformation E. So we say that 'A' is preserved by the *group* of transformations consisting of $\{R_V, \mathrm{E}\}$.

Problem 1: Determine the group of transformations that preserve the other letters of the alphabet. Notice in Fig. 21.1b that the letter 'p' is preserved by only the identity, E. As a result we say that 'p' is *asymmetric*, i.e., it has no symmetries other than the trivial one — the identity.

21.2. Symmetry of a Square

There are eight operations that change the position of points within the square but map the entire square onto itself. These operations are illustrated in Fig. 21.2 symbolized by:

$$E, S_{90}, S_{180}, S_{270}, R_V, R_H, D_+, D_-$$

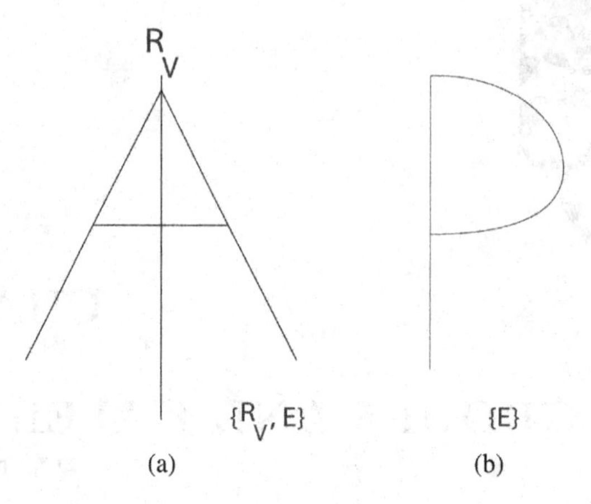

Fig. 21.1. Symmetry of the letter 'A' and 'P'.

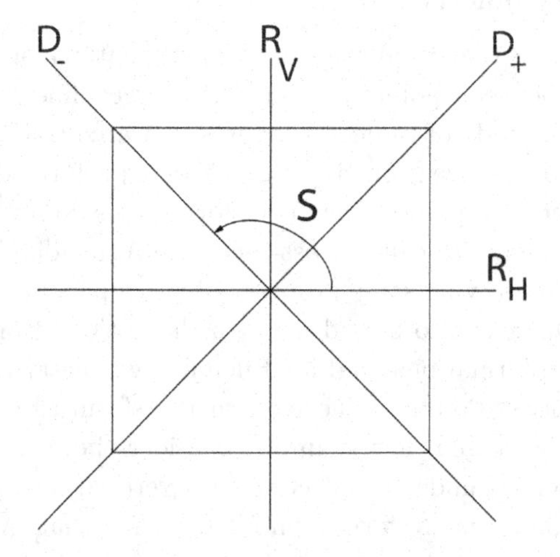

Fig. 21.2. The symmetries of a square.

where,

E is the identity transformation (leaves the square unchanged)
S_{90} is a counterclockwise quarter-turn
S_{180} is a half-turn
S_{270} is a counterclockwise three-quarter turn (or 90 deg. clockwise rotation)
R_V is a vertical flip
R_H is a horizontal flip
D_+ is a flip in the positive sloped diagonal of the square
D_- is a flip in the negative sloped diagonal of the square.

Remark: Since a flip exchanges a pair of regions, i.e., left half of the square with right half or top half with bottom half of the square, it is equivalent to a reflection.

Table 21.1. A multiplication table for the symmetries of a square.

Transformation	Orientation of p
E	p
S_{90}	
S_{180}	d
S_{270}	
R_V	q
R_H	
D_+	
D_-	

Cut a square from a sheet of blank white paper. Draw a large 'p' in the center of the square, and trace the p on the back of the square so that it takes the form of the reflection of the p. Then subject the p to each of the above operations. For example, the identity operation maps p to itself, R_V maps p to q, while S_{180} maps p to d. Draw the position of the p for each of the other five operations, and associate each operation with a different position of the p. Record your results in Table 21.1.

21.3. Multiplications of Symmetry Transformations

If you carry out transformation T_1, and follow it by transformation T_2, since the square is preserved by each of the operations, the result could have been carried out in a single step by one of the above operations, say transformation T_3. We denote this by the symbolic statement:

$$T_2 \circ T_1 = T_3.$$

In what follows, we will leave out the small circle and write this as,

$$T_2 T_1 = T_3.$$

This appears to be a kind of "multiplication" of transformations, and I will refer to it as a multiplication.

Remark: It is important to remember that the first operation is listed on the right, followed by the second operation on the left.

Problem 2: Complete the following "multiplication table" for the symmetry group of a square. This symmetry group, $D_4 = \{E, S_{90}, S_{180}, S_{270}, R_V, R_H, D_+, D_-\}$, is referred to as the dihedral or kaleidoscope group with four mirrors or D_4. It can be helpful to begin with a p facing forward on the face of the square, carry out operation 1, then perform operation 2 and record the resulting orientation of the p in Table 21.2. After doing this you can go to Table 21.1 to correlate this position with the appropriate transformation. For example, a vertical reflection followed by a horizontal reflection,

Table 21.2. Multiplication table for the symmetries of a square.

Op 1 \ Op 2	E	S_{90}	S_{180}	S_{270}	R_V	R_H	D_+	D_-
E								
S_{90}								
S_{180}								
S_{270}								
R_V						S_{180}		
R_H								
D_+								
D_-								

i.e., $R_H \circ R_V$, takes p into d so that it follows that $R_H \circ R_V = S_{180}$. In this way you can complete the Table 21.2. Try to do this on your own before looking at the result in Appendix 21A.

When you open up the 45 deg. snowflake pattern from Chap. 20, notice that the pattern replicates eight times in different orientations just as the p was replicated in eight orientations in Table 21.1. This is shown in Fig. 21.3. In other words, for the snowflake, the pattern that you cut into the folded wedge takes the place of the p. This is how all symmetries are created. Just take a fundamental pattern (the pattern in the 45 deg. wedge for the case of the snowflake) and transform it by each of the operations of the group that keeps some pattern (e.g., the square in the case of the snowflake) unchanged or invariant.

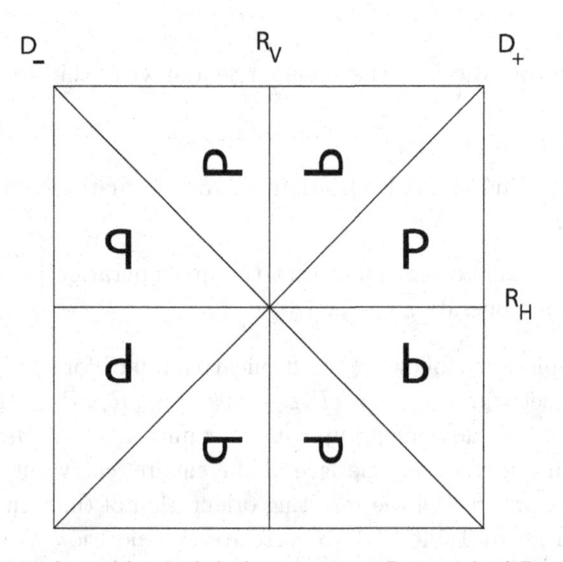

Fig. 21.3. P is transformed by the kaleidoscope group D_4 to eight different orientations.

21.4. Symmetries and Mirrors

You have already learned in Chap. 19 that isometries can be carried out using only mirror reflections. This is the principle of the kaleidoscope. Now take a pair of mirrors and place them at a 45-deg. angle. Draw a large p on a piece of paper between the two mirrors and notice how the p is transformed into a set of seven additional p's in the mirrors. Notice that the set of eight p's are exactly the eight transformations of the letter p in Table 21.1 and the pattern is identical to the one in Fig. 21.3. This means that your design can only be made in one-eighth of the square as was done in the construction of the snowflake pattern of Chap. 20. The symmetries of the square will spread it to the remaining sectors of the square as in Fig. 21.3. We say that the 45-deg. sector of the square is the *fundamental domain* of kaleidoscope group, D_4. You can experiment with other dihedral symmetries, D_n, by placing the mirrors at the angles: $180/n$.

Problem 3: Create a multiplication table for the group of isometries of a rectangle, D_2 (the kaleidoscope group with two mirrors), i.e., the set of isometries that keep a rectangle invariant. What is its fundamental domain?

21.5. Symmetry Groups as an Algebraic System

If you study Table 21.2, you will see that it has a life of its own. What do you observe about Table 21.2? Notice that each element of the group appears once in each row and column. Also notice from Table 21.2 that the *associative law* of algebra works for transformations, i.e.,

$$T_3 \circ (T_2 \circ T_1) = (T_3 \circ T_2) \circ T_1. \tag{21.1}$$

Try this out for several examples from Table 21.2, e.g., $(S_{90}\, R_V)D_+ = S_{90}(R_V\, D_+)$ or $(D_+\, R_V)R_H = D_+(R_V\, R_H)$, etc.

Remark: Since the grouping of the terms in the compound multiplication of transformations does not matter, we can eliminate the parentheses. However, order does matter since, in general, $T_2 \circ T_1 \neq T_1 \circ T_2$, i.e., the *commutative law* does not always hold.

Problem 4: Find a pair of transformations in Table 21.2 for which the commutative law does not hold.

As a result of the associative law, $D_+\, S_{90}$ can be found algebraically. Observe in Fig. 21.3 that a horizontal reflection followed by a reflection in the positive leaning diagonal results in a 90-deg. counterclockwise rotation, i.e., $S_{90} = D_+\, R_H$ because R_H and D_+ are mirrors inclined at 45 deg. which results in a 90-deg. rotation. So,

$$D_+\, S_{90} = D_+\, D_+\, R_H = R_H$$

since $D_+\, D_+ = E$. (Why?) We now have an algebraic way to compute the terms in Table 21.2 without having to physically manipulate the square. Use this approach to find:

$$R_H\, D_+ = \_\_\_, \quad S_{90}\, R_H = \_\_\_, \quad \text{and} \quad D_-\, S_{180} = \_\_\_.$$

In the number system, 0 is the identity element under addition since 0 added to any number results in the same number. The E transformation is the identity element in Table 21.2. Also, each number has an additive inverse so that if you add the inverse to the number you get back the identity, e.g., $2+(-2) = 0$ so -2 is the additive inverse of 2.

In general, the inverse of a transformation T, denoted by T^{-1}, has the following algebraic property,

$$TT^{-1} = E = T^{-1}T. \tag{21.2}$$

What is the inverse of each of the eight symmetries of the square? These can be immediately read off from Table 21.2 (or Table 21A). For example, $S_{90}^{-1} = S_{270}$. Can you see this?

Once you know the inverses of each element of the group you can use this information to solve transformation equations. For example, solve for transformation X when,

$$XS_{90} = D_-. \tag{21.3}$$

Of course this equation can be solved for X directly by looking at Table 20.2 (or Table 21A). However, it can also be solved by multiplying both sides of Eq. 21.3 by S_{90}^{-1}, i.e.,

$$XS_{90}S_{90}^{-1} = D_-S_{90}^{-1}.$$

From Eqs. 21.2,

$$XE = X = D_-S_{90}^{-1}.$$

Since $S_{90}^{-1} = S_{270}$, by reading the result from Table 21.2,

$$X = D_-S_{270} = R_H.$$

Compare the solution of Eq. (21.3) to solving for in x.

$$5x = 8.$$

Solution:

$$5^{-1}5x = 5^{-1}8$$

$$1x = \frac{1}{5} \cdot 8$$

$$x = \frac{8}{5}$$

This application of groups to solving equations reveals the idea behind how all equations of high school algebra are solved.

Problems:

5. Show that $S_{180}(R_HD_-) = (S_{180}R_H)D_-$.
6. Solve for X where $XS_{270} = D_+$.
7. Solve for X where $R_VX = S_{90}$.

21.6. Group Theory

The concept of a *group* lies at the basis of mathematics and its applications. Groups have applications to quantum mechanics, crystallography, design and even music. Everything that you have learned about arithmetic and algebra has its roots and rationale in the theory of groups.

For a set of elements, along with a well-defined operation of multiplication, to qualify as a group, in a mathematical sense, it must have the following properties:

a. *Closure*: The product of two elements from the group must itself lie in the group.
b. The *associate law* must hold for all elements in the group.
c. There must be an *identity element.*
d. Each element of the group must have an *inverse* element also in the group.

Clearly, each set of kaleidoscope symmetries constitutes a group.

Problem: Given an algebraic equation, sometimes there is a single solution, and sometimes there are more than one solution. For example, $2x+1=7$ has the single solution $x=3$ whereas $x^2=4$ has two solutions, $x=2$ and $x=-2$. Consider the equations,

$$XA = B \quad \text{and} \quad AX = B$$

where A, B and X are three elements of a group. Show that each of these equations has a single solution by using the properties of a group listed above. This explains why every row and column of the multiplication table of a group has each element of the group listed once. (Why?)

Appendix 21A. Multiplication Table for the Symmetries of a Square

Table 21A.1. Multiplication table for the symmetries of a square.

Op 1 \ Op 2	E	S_{90}	S_{180}	S_{270}	R_V	R_H	D_+	D_-
E	E	S_{90}	S_{180}	S_{270}	R_V	R_H	D_+	D_-
S_{90}	S_{90}	S_{180}	S_{270}	E	D_+	D_-	R_H	R_V
S_{180}	S_{180}	S_{270}	E	S_{90}	R_H	R_V	D_-	D_+
S_{270}	S_{270}	E	S_{90}	S_{180}	D_-	D_+	R_V	R_H
R_V	R_V	D_-	R_H	D_+	E	S_{180}	S_{270}	S_{90}
R_H	R_H	D_+	R_V	D_-	S_{180}	E	S_{90}	S_{270}
D_+	D_+	R_V	D_-	R_H	S_{90}	S_{270}	E	S_{180}
D_-	D_-	R_H	D_+	R_V	S_{270}	S_{90}	S_{180}	E

C.6. Group Theory

[illegible]

a. Closure: [illegible]
b. The associative law should [illegible]
c. There should be an identity element.
d. Each member of the group must have an inverse [illegible]

[illegible]

Appendix [illegible] Multiplication Table [illegible]

[illegible]

CHAPTER 22

FRIEZE PATTERNS

22.1. Introduction

Frieze patterns are symmetric patterns along a horizontal strip. They can be seen as the ornamental patterns decorating the edge of buildings. Mathematically speaking, Frieze patterns extend to infinity in both directions. Of course, in practice, only a finite segment of the infinite pattern is used. Figure 22.1 illustrates seven examples of Frieze symmetry patterns. These seven examples represent one each of the seven possible classes of Frieze patterns. Schematic drawings of these seven patterns are shown in Table 22.1, once again using the letter p. It can be proven that there can be no other possibilities [Mey].

22.2. Explanation of the Seven Frieze Patterns

If each of these infinite patterns is subjected to certain isometries, the points within the pattern change their locations but the entire pattern is unchanged, or as mathematicians say, they are *invariant*. For example, each of the seven classes of Frieze patterns are invariant under translations along the horizontal strip. Let's examine these seven classes of Frieze patterns as shown in Table 22.1.

Pattern 1, F_1, is generated by only translations. Starting with a single 'p' and translating it to the right and to the left by a given translation vector results in the entire pattern. The translation vector specifies the distance from one p to the next. Only within the region from one p to the next are you permitted to create a pattern, in this case a single p. The remaining patterns are replications of the first under translation. This region between p's is referred to as the *fundamental domain*. We saw something similar in our construction of snowflakes in Chap. 20. For the snowflake corresponding to the symmetry of a square, you were permitted to create a pattern in only one-eighth of the square.

Pattern 2, F_1^3. Notice that the p is glide reflected across the center line to a 'b'. A second glide reflection brings you back to the p. The entire pattern is generated by glide reflections, and the fundamental region is the region beneath the centerline and between p and half the distance to the next p.

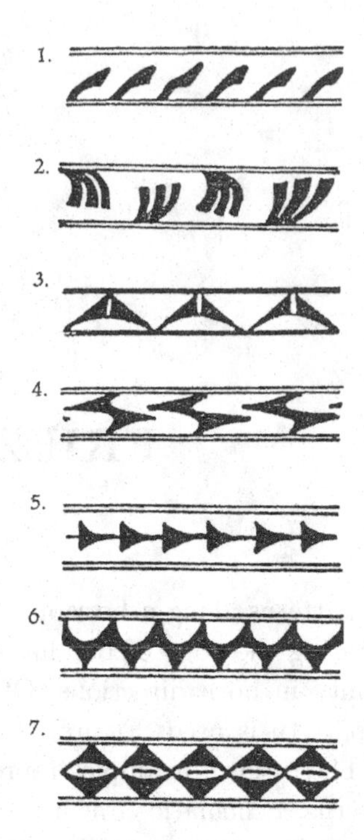

Fig. 22.1. Patterns illustrating the seven frieze symmetries.

Table 22.1. Illustration of the seven Frieze patterns.

F_1	p p p p p	1 translation
F_1^3	b b p p p	1 glide reflection
F_1^2	q¦p ¦ qp qp	2 reflections
F_2	d•p • dp dp	2 half turns
F_2^2	q¦p • db qp	1 reflection and 1 half turn
F_1^1	b b b b b p p p p p	1 translation and 1 reflection
F_2^1	d¦b ¦ db db db q¦p ¦ qp qp qp	3 reflections

Pattern 3, F_1^2, has two vertical mirrors. A p reflects to its mirror image q in the first mirror. The pair qp then translates through twice the distance between the mirrors. In this case the fundamental domain is the region between the two mirrors. Repeated reflections in this pair of vertical mirrors generate the translations that recreate the entire pattern.

Pattern 4, F_2, is generated by half-turns in two given fixed points, A and B. Here p is rotated to d in the first fixed point A, and the result rotates by a half turn in the second fixed point B to d p. Successive rotations in the two fixed points generate a translation as we saw in Problem 5 of Sec. 19.5, i.e., $S_{180}^B S_{180}^A = T_{2(B-A),0}$. The fundamental domain is the region between the two fixed points of rotation.

Pattern 5, F_2^2, is generated by a reflection in a vertical mirror between p and q, followed by a rotation by a half-turn about the given fixed point to d b. Since the vertical mirror is also rotated through 180 deg., the pair of mirrors translate p through twice the distance between the mirrors to the next image. The fundamental domain is the region between the vertical mirror and the fixed point of the half-turn.

Pattern 6, F_1^1, is generated by a translation between two successive p's and a reflection across a mirror on the centerline. The fundamental domain is the region below the centerline between two successive p's. Beginning with p, successive translations and reflections result in the entire pattern.

Pattern 7, F_2^1, is generated by two vertical mirrors and one horizontal mirror. A p placed below the horizontal mirror and between the vertical mirrors, reflects to q after which the combination qp reflects in the horizontal mirror to $\begin{matrix} db \\ qp \end{matrix}$ after which this configuration is translated successively through twice the distance between the vertical mirrors. The fundamental region is the region between the pair of vertical mirrors and the centerline, containing the original p. You can only create a pattern here after which successive application of the three reflections result in the entire pattern. Note that again, translations are generated by reflections in the two vertical mirrors.

22.3. Identifying Frieze Patterns

So we see that Frieze patterns display translation, rotation in a half-turn, reflection in vertical or horizontal mirrors, and glide reflections. It is sometimes difficult to classify Frieze patterns such as the ones in Fig. 22.1. The flowchart in Fig. 22.2 should help you in this task. In this flowchart, when it says 'reflection in a center' it refers to a reflection in a mirror along the horizontal or centerline. Otherwise the reflections are in vertical mirrors. A helpful hint towards identifying half-turns is to turn the pattern upside-down. If the pattern is unchanged, then there is a half-turn somewhere in the pattern.

Problem 1: Use the flowchart in Fig. 22.2 to classify the seven patterns in Fig. 22.1 and indicate their fundamental domains.

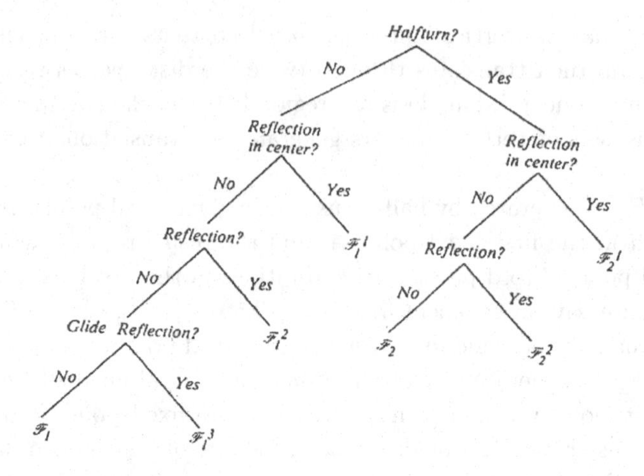

Fig. 22.2. A tree to aid in classifying the Frieze symmetry group.

Problem 2: Either take a photo of several patterns from the edge of buildings, or find patterns in books or magazines of architecture and design, and use Fig. 22.2 to classify the patterns. For each pattern, indicate the fundamental domain.

For the next problem, you will need two soap erasers and a stamp pad.

Problem 3: Construct your own example of each of the seven Frieze patterns by using two soap erasers and an Exacto knife. Carve a pattern in one eraser and its mirror image in the other. It is important that the patterns that you carve should be asymmetric in both the horizontal and vertical directions. Then, use a stamp pad to stamp out each of the seven Frieze patterns similar to the patterns in Figs. 22.1 and 22.2. Your pattern on the first eraser now takes the place of the p's in Table 22.1. Half-turn rotations can be carried out with the first eraser since half-turns are proper isometries. However, reflections require you to use the mirror image pattern since reflections are improper, and therefore it requires you to use the second eraser pattern. In this way, the p's in all of its various orientations for the patterns in Table 22.1 are rendered as juxtapositions of the two erasers.

22.4. Additional Problems

Many additional Frieze patterns are shown in Fig. 22.3a. Try your hand at classifying some of them. Word patterns corresponding to the seven Frieze groups are shown Fig. 22.3b. Can you identify the groups?

(a)

ME COOK·HE COOK·WE COOK ·HE C
SANTA CLAUS SANTA CLAUS SA
HOHOHOHOHOHOHOHOHOHOHOHOHO
MTMTMTMTMTMTMTMTMTMTMTMTMT
NOON NOON NOON NOON NOON NO
WOMOWOMOWOMOWOMOWOMOWOMO
HOT TOMATO HOT TOMATO HO

(b)

Fig. 22.3. Examples of Frieze symmetries (a) patterns; (b) words.

CHAPTER 23

AN INTRODUCTION TO FRACTALS

23.1. Introduction

Euclidean geometry has had a major impact on the cultural history of the world. Not only mathematics, but art, architecture, and the natural sciences have utilized the elements of Euclidean geometry or its generalizations to projective and non-Euclidean geometries. However, by its nature, Euclidean geometry is more suitable to describe the ordered aspects of phenomena and the artifacts of civilization rather than as a tool to describe the chaotic forms that occur in nature. For example, the concept of point, line, and plane which serve as the primary elements of Euclidean geometry, are acceptable as models of the featureless particles of physics, the horizon line of a painting, or the façade of a building. On the other hand, the usual geometries are inadequate to express the geometry of cloud formation, the turbulence of a flowing stream, the pattern of lightening bolts, the branching of trees and alveoli of the lungs, the variability of the stock market, or the configuration of coastlines.

In the early 1950's, mathematician Benoit Mandelbrot, aware of work done half a century before, rediscovered geometrical structures suitable for describing these irregular sets of points, curves, and surfaces from the natural world. He coined the term *fractal* for these entities and invented a new branch of mathematics to deal with them. The key to understanding fractals and their applications to the natural world lies in their embodiment of self-similarity, the endpoint of an infinite process. Unlike calculus which was invented by Newton and Leibnitz in the 17th century to deal with the variability and change observed in dynamic systems such as the motion of the planets and mechanical devices using curves that were mostly smooth, fractals are generally depicted by structures that are nowhere smooth. In fact, many of the so-called pathological curves which were discovered by mathematicians studying the foundations of calculus and then banished from its further study became the starting points for this new discipline of fractals [Kap1], [Kap2], [Fra], [Man], [Pei].

We have already seen in Sec. 3.3, beginning with a line segment the development of a simple fractal that is nowhere smooth. Let us see how other fractals are generated.

23.2. The Koch Snowflake

The *Koch snowflake* is created by the following steps:

1. Begin with a line segment as shown in Fig. 23.1a.
2. Subdivide the line segment into three equal parts and replace the middle third by a pair of edges of the same length to form a tentlike structure with four equal line segments as shown in Fig. 23.1b.
3. Apply Steps 1 and 2 to each of the resulting four line segments (see Fig. 23.1c).
4. Continue this process to later stages in the development of the Koch snowflake. A later stage is shown in Fig. 23.1d.
5. The same process could also be carried out beginning with a triangle as shown in Fig. 23.1.

Remark: It turns out that at the end of this infinite process, the snowflake will be infinite in length and nowhere smooth. In fact, the distance between any two points on the snowflake, no matter how close, will be infinite. Also, observe in Fig. 23.1d that each segment of the fractal appears like the whole except at a smaller scale. This self-similarity becomes exact at the infinite stage. We will find this self-similarity ubiquitous to fractals.

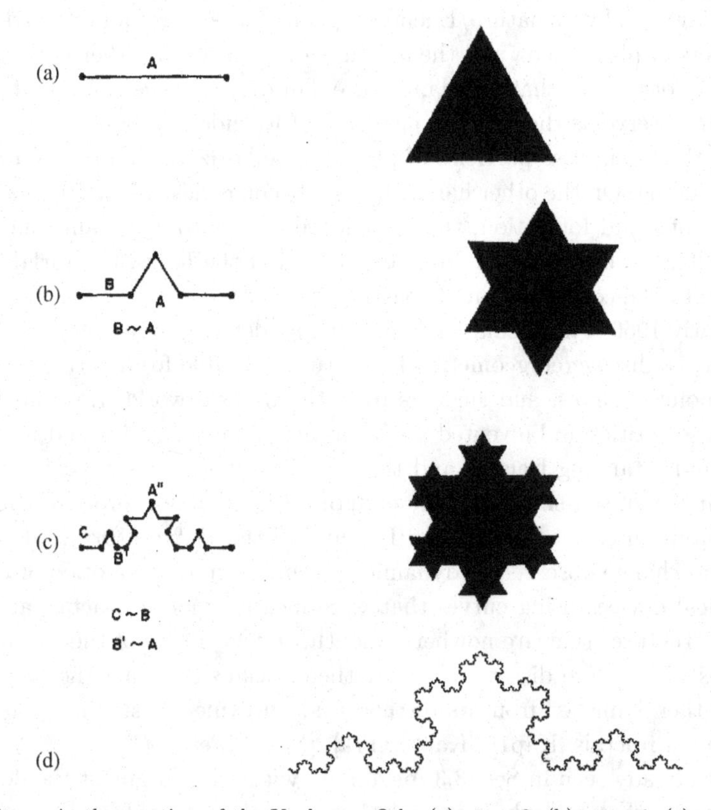

Fig. 23.1. Stages in the creation of the Koch snowflake (a) stage 0; (b) stage 1; (c) stage 2; (d) an advanced stage.

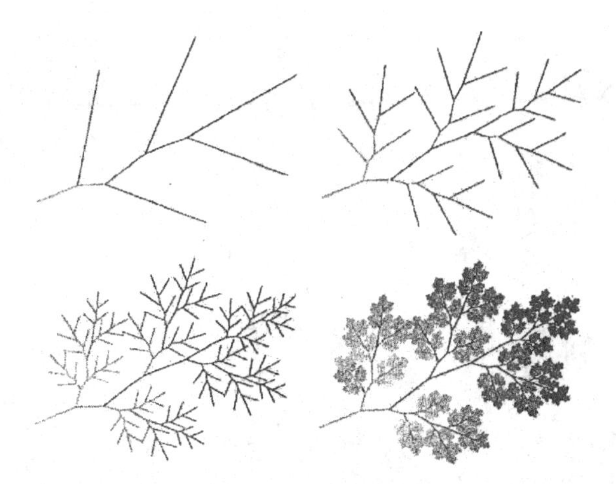

Fig. 23.2. Three stage to the creation of a fractal tree branch leading to an advanced stage.

23.3. A Fractal Tree

1. Begin with a rudimentary tree branch with five branch tips (see Fig. 23.2a).
2. Replace each branch tip by an exact miniature of the original branch (see Fig. 23.2b). The tree now contains 25 branches.
3. Replace each of the 25 branches again with a miniature of the original so that Fig. 23.2c possesses 125 branches.
4. Repeat this process infinitely to obtain the fractal tree branch shown in Fig. 23.2d.

Remark: In the early stages of this process, the tree branch is not self-similar. However, in advanced stages, its self-similarity becomes evident, and it is fully manifested in the infinite stage of its development. Of course, we can never reach this infinite step, and so we must stop the process at a sufficiently late stage.

Remark: Notice that the whole tree is self-similar to a single branch or even a single leaf.

23.4. A Moonscape

Annalisa Crannell and Marc Frantz [Fra] put forth the following fractal simulation of a moonscape.

1. Begin with a circular "crater" (see Fig. 23.3a) placed in a square.
2. Randomly add to the square eight circular "craters" of width $\frac{1}{3}$ the original (see Fig. 23.3b) to the square.
3. Repeat this process again with the addition of 64 craters of size $\frac{1}{9}$ the original (see Fig. 23.3c).
4. In the following two stages, 512 and 4096 craters are added shown in Figs. 23.3d and 23.3e.
5. A photograph of the Moon is shown in Fig. 23.3f.

Each segment of the square looks very much like every other segment, again reflecting the self-similarity of the moonscape. After all, outside influences that lead to the

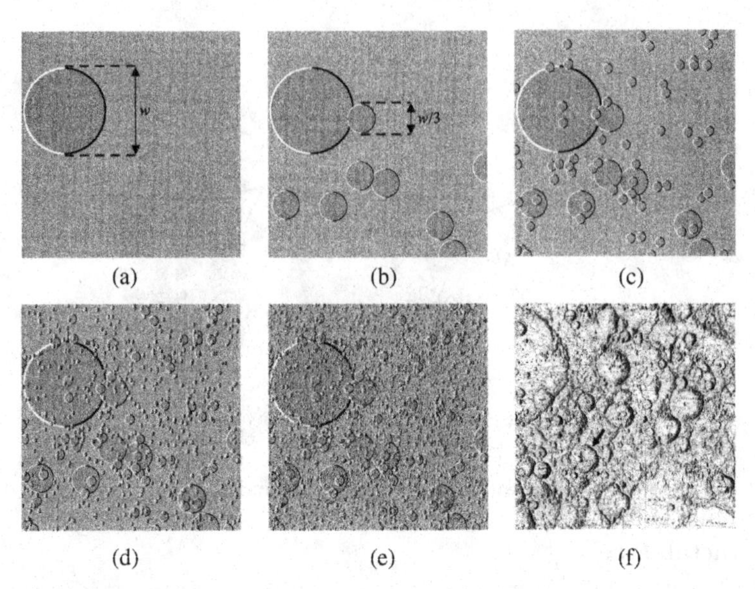

Fig. 23.3. Mimicking the construction of a moonscape: (a) a giant crater of width w; (b) 8 "large" randomly-placed craters, each 1/3 the size of the giant one; (c) $8^2 = 64$"medium" craters, each 1/3 the size of the large ones; (d) $8^3 = 512$"small" craters, each 1/3 the width of the medium ones; (e) $8^4 = 4096$ "tiny" craters, each 1/3 the width of the small craters; (f) An actual moonscape.

moonscape such as meteor impacts, act in a similar way within each sub-portion of the square and should, therefore, result in a scape with similar appearance.

23.5. A Cauliflower

Purchase a head of cauliflower from your local supermarket. Observe how each of the florets appears to be self-similar to the whole cauliflower. We can artificially reproduce a facsimile of the cauliflower by the process shown in Fig. 23.4, which needs no further explanation.

23.6. A Fractal Wallhanging

What we have seen for the Koch snowflake, the tree branch, the moonscape and the cauliflower are representative of countless other fractals. We will now present a rich source of fractal designs known as the Iterative Function System [Fra]. To introduce this system, a fractal will be generated in the same manner as the ones presented in Secs. 23.2–23.5 and then scaled up to create a wallhanging.

1. Begin with a black square.
2. Iteration 1: Scale the square down to half of its size, and

   a. place an exact copy in the lower right-hand quarter of the original square,
   b. rotate the black square counterclockwise by 90 deg. and place the result in the upper left-hand quarter of the original square, and

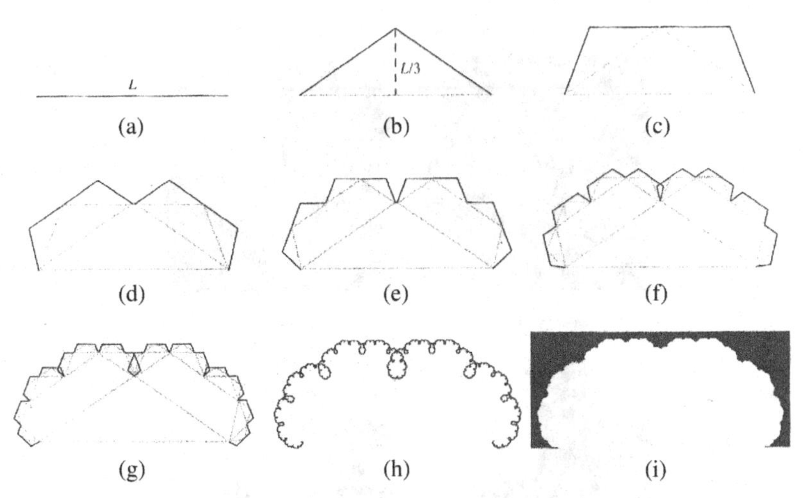

Fig. 23.4. Steps in the creation of a "cauliflower" (a)–(g) and a computer rendering; (h) after many iterations. Part (i) shows the region above the cauliflower in black.

c. rotate by a half-turn (180 deg.) and place the result in the lower left-hand quarter of the original square.
d. leave the upper right-hand quarter of the original square blank.

The resulting L-shaped figure resides in the original square frame.

3. Iteration 2: Take the result of Iteration 1, scale it to half its size, and perform the same three operations as in Step 2,

   a. the identity transformation — lower right-hand quarter,
   b. a counterclockwise rotation of 90 deg. — upper left-hand quarter,
   c. a rotation of 180 deg. — lower left-hand quarter,

   with the three results placed into the original square frame.
4. Iterations 3,4 and 5 are carried out in the same way.
5. Iteration 5 is reasonably close to the infinite stage shown in Fig. 23.6.
6. Now begin to enlarge the fractal by taking three copies of the infinite fractal (actually three copies of a late stage). This time, instead of scaling down by half, we scale up by twice its size by the following procedure:

   a. Take three copies of a 2 in. by 2 in. picture of the fractal shown in Fig. 23.6. Place one copy of the fractal in the lower right-hand quarter of a 4 in. × 4 in. scaled up square.
   b. Rotate a second copy of the fractal counterclockwise by 90 deg. and place it in the upper left-hand quarter of the 4 in. by 4 in. square.
   c. Give a third copy a half-turn, and place the result in the lower left-hand quarter of the 4 in. by 4 in. square.
   d. Place an empty 2 in. by 2 in. square in the upper right-hand quarter of the 4 in. by 4 in. square.

7. In Step 6, you have created a fractal scaled up twice the size of the one in Fig. 23.6. Take three copies of the scaled up fractal and carry out the same procedure as Step 5

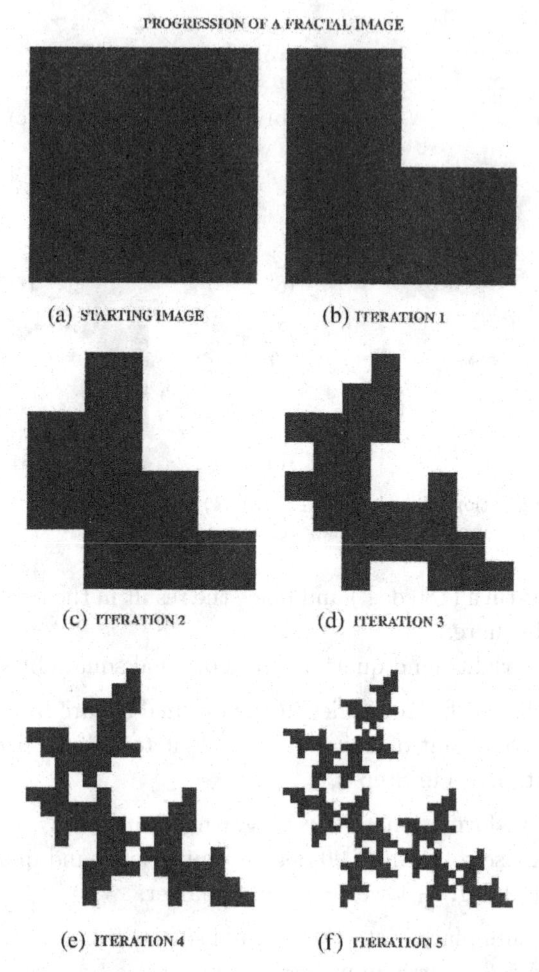

Fig. 23.5. Five stages in the development of a fractal from the Iterative Function system beginning with a black square.

to recreate the fractal at twice again the size, i.e., four times the size of the original fractal or 8 in. by 8 in. At this stage you will need 9 copies of the fractal shown in Fig. 23.6.

8. If you repeat this procedure two or three more times, you will have a greatly enlarged 32×32 or 64×64 fractal wall-hanging.

Remark: To create this wallhanging, we limit ourselves to proper transformations of the square, namely, the identity transformation, quarter-turn, half-turn, and three-quarter turn. In principle, we could have considered improper transformations such as reflections in the vertical or horizontal, or reflections in the diagonals, i.e., improper isometries. However, these would have been difficult to physically manipulate in the iteration process.

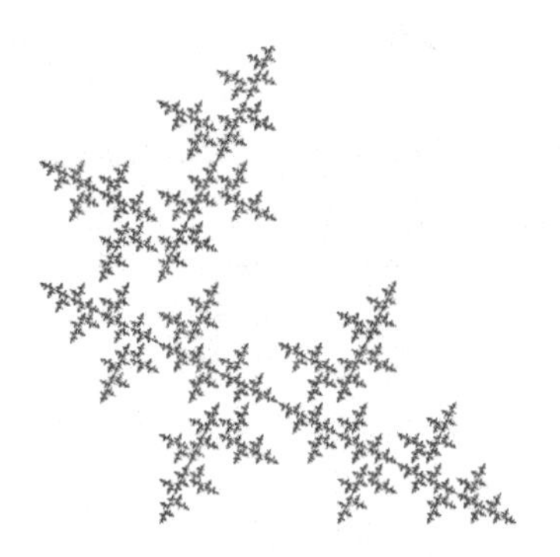

Fig. 23.6. A late stage of the fractal.

Remark: The creation of this fractal began with a black square. It is quite amazing that any starting pattern within the original square frame, including the letter 'p', or even a scribble, would result, at the end of the day, in the same result.

23.7. Finding the "DNA" of an Iterative Function System

Working backwards from the fractal in Sec. 23.6, we see that it is characterized by the three isometries leading to its formation. I shall refer to this as the DNA code of the fractal, i.e., the DNA of the fractal in Fig. 23.6 is (E, S_{90}, S_{180}) corresponding to the sub-squares: (lower right, upper left, lower left). A family of fractals can be constructed related to the one given in Fig. 23.6. Each of the fractals in this family, as for the one in Fig. 23.6, can be subdivided into three square sectors embedded in the original square frame along with an empty square on the upper right quadrant. Each sub-square is a scaled copy of the entire fractal subjected to one of the eight isometries of the square. There are eight possibilities for each sub-square; so there will be $8 \times 8 \times 8 = 512$ different fractals. Four of these are shown in Fig. 23.7.

Remark: The pattern within each sub-square is a transformed version of the entire fractal. Likewise, each sub-pattern can also be subdivided into even smaller sub-patterns which are similar to the whole pattern. Therefore, these fractals are self-similar at many scales.

Problem: Identify the DNA codes of the four fractals in Fig. 23.7.

The trick is to find the correct isometries which transform the original fractal to the patterns within each sub-square. The next section will make this task easier.

23.8. Symmetry Finder

Finding the DNA of the fractal can be daunting, particularly when the isometry is improper. We can simplify this search by using a symmetry finder.

1. A symmetry finder begins with the pair of squares shown in Fig. 23.8.
2. Place a small version of the fractal in the left-hand square, and place its mirror image in the right-hand square.

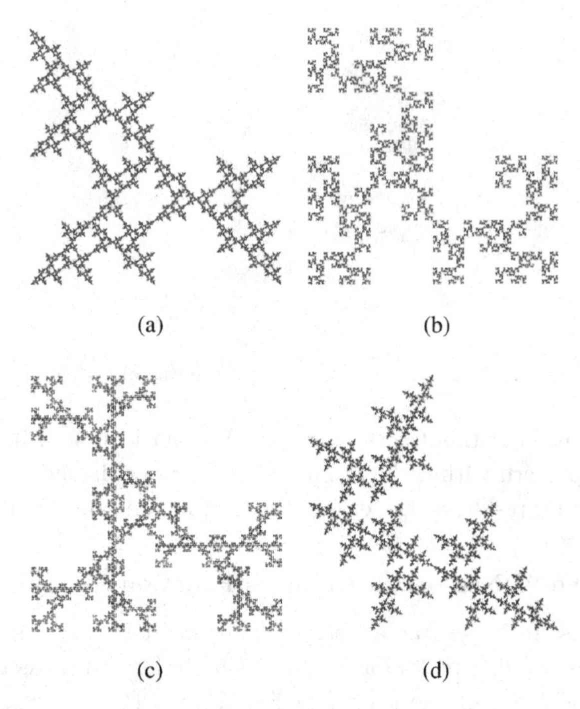

Fig. 23.7. Four fractals from the Iterative Function System.

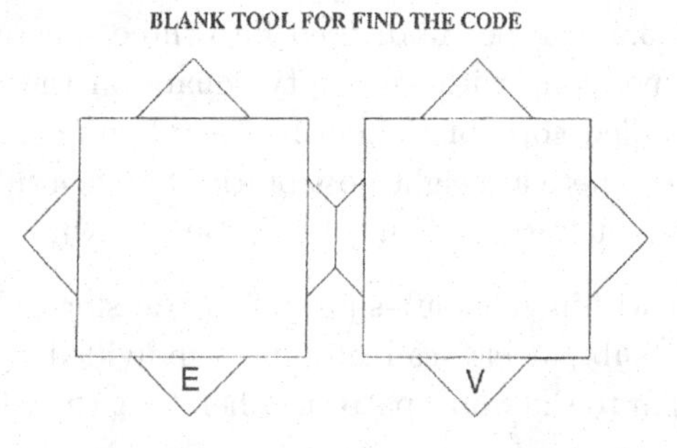

Fig. 23.8. The symmetry finder.

3. At the bottom of the left square there is an E symbolizing the identity, and at the bottom of the right square is placed a V which is a shorthand for R_V symbolizing a vertical reflection.
4. Cut out the symmetry finder and fold it so that E is on the front and V is on the back.
5. Notice that it has 8 orientations corresponding to each of the 8 isometries of the square (see Chap. 21). Turn the square face up, and then subject it to each of the 8 isometries.
6. You will notice that one of the six blank tabs (two tabs are already marked with E and V) is now at the bottom. Write the name of the appropriate isometry on that tab. For example, if you give the square a flip about the horizontal axis (equivalent

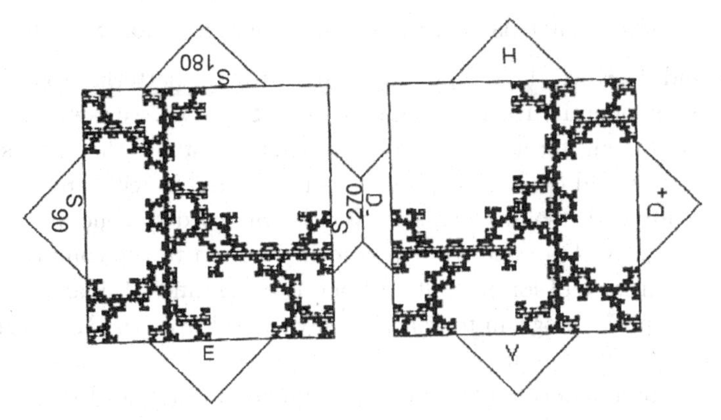

Fig. 23.9. The symmetry finder with the fractal on the left hand square labeled E and its mirror image on the right hand square labeled V.

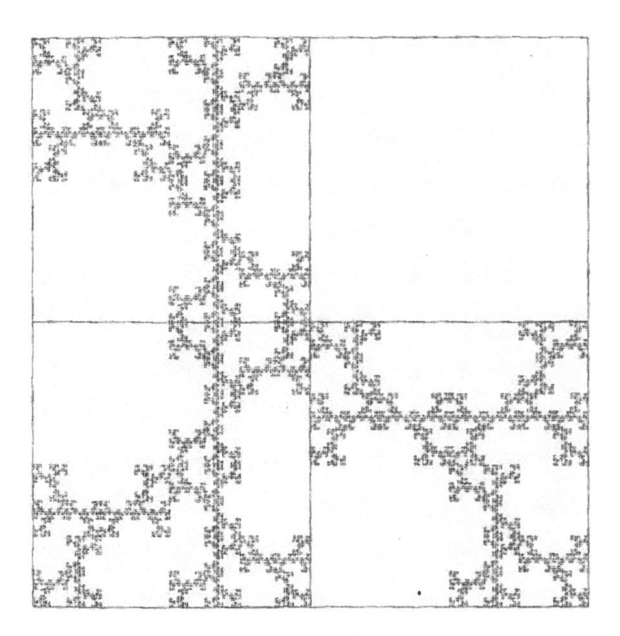

Fig. 23.10. The fractal from Fig. 23.7(c) subdivided into three sectors.

to a reflection in the horizontal) place H as a shorthand for R_H at the bottom of the square.

7. Now you are ready to twist and turn the symmetry finder until you see the image of the fractal pattern that you are looking for. The identification of the isometry will be found by reading the tab at the bottom.
8. In this way you can identify the DNA codes that lead to the patterns in Fig. 23.7. To help you get started, cut out the symmetry finder in Fig. 23.9 which is the same pattern as in Fig. 23.10 with lines drawn demarcating the sub-squares also shown in Fig. 23.7c. Note that the eight isometries are marked on the tabs. Apply the above procedure to find the code.

23.9. Construction Fractals from the Iterative Function System

In Secs. 23.7 and 23.8, we showed how to analyze fractal patterns from the Iterative Function System by identifying their DNA codes. In Sec. 23.6, we saw that the creation of fractal patterns is an iterative procedure in which we apply the same set of transformation rules over and over. This process can be initiated with any pattern drawn in a square including the solid black square that was used to generate the fractal in Fig. 23.6. The Iterative Function System is ideally suited to the computer, and computers are an essential tool for generating fractals of all kinds. Ljiljana Radovic [Rad] has created a computer program to generate any fractal from the IFS beginning with the DNA code and a black square.

The next chapter is devoted to studying the mathematics needed to communicate with a computer in order to carry out isometries and create fractals as Radovic did with the IFS system.

CHAPTER 24

ISOMETRIES AND MATRICES

24.1. Introduction

In Chaps 18 and 19, we used compass, straightedge and protractor to carry out isometries and their products. In Chap. 23, we went further and used isometries and scaling to create fractals that were self-similar at all scales. In this chapter, we introduce the mathematics that enables the computer to generate isometries, scaling, and fractals. The same mathematical tools will also be applied to projective geometry. Mathematical structures called *vectors* and *matrices* are the necessary tools to accomplish these tasks. The theory behind vectors and matrices lies in the realm *of linear algebra.* This is a huge area of study. However, we will be able to satisfy our needs with only a few ideas from this discipline. Isometries are one of many so-called *linear transformations.* Vectors were described in Chap. 15 while an introduction to matrices is found in Appendix 24A.

24.2. Matrices for Rotations and Reflections in the Plane

Matrices to carry out linear transformations can be constructed by knowing how two special vectors, $\left[\begin{smallmatrix}1\\0\end{smallmatrix}\right]$ and $\left[\begin{smallmatrix}0\\1\end{smallmatrix}\right]$ transform. These vectors are shown in an (x, y)-coordinate system in Fig. 24.1. We showed in Sec. 15.10 that vector $\left[\begin{smallmatrix}a\\b\end{smallmatrix}\right]$ can be represented by an arrow drawn from the origin of an (x, y)-coordinate system to the point (a, b) also shown in Fig. 24.1 for the vector, (2, 1).

Note that for convenience, I will sometimes use the notation $\left[\begin{smallmatrix}a\\b\end{smallmatrix}\right] \equiv (a, b)$.

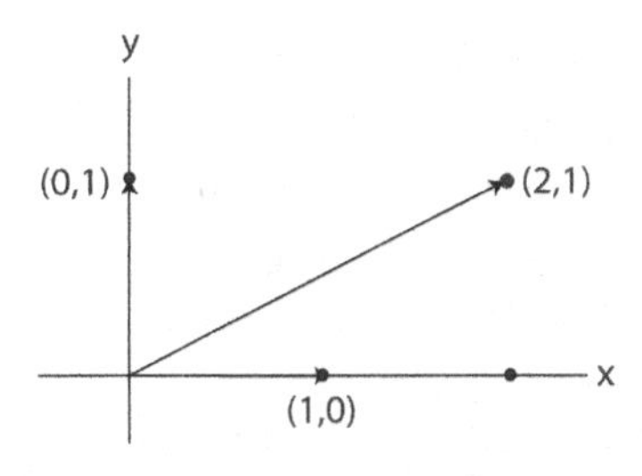

Fig. 24.1. A vector shown in an x, y-coordinate system.

a. For example, consider a 90-deg. counterclockwise rotation, i.e, S_{90}

$$(1,0) \rightarrow (0,1)$$
$$(0,1) \rightarrow (-1,0).$$

The result of the 1st transformation is the 1st column of the matrix, while the result of the 2nd transformation is the 2nd column, i.e.

$$S_{90} = \begin{bmatrix} 0 & -1 \\ 1 & 0 \end{bmatrix}. \tag{24.1}$$

We can use this matrix to transform any position vector (point) in the plane by matrix multiplication (see Appendix 24A). For example, (2, 1) and (0, 0) transform as follows:

$$\begin{bmatrix} 0 & -1 \\ 1 & 0 \end{bmatrix} \begin{bmatrix} 2 \\ 1 \end{bmatrix} = \begin{bmatrix} -1 \\ 2 \end{bmatrix}$$
$$\begin{bmatrix} 0 \\ 0 \end{bmatrix} = \begin{bmatrix} 0 \\ 0 \end{bmatrix}$$

$S_{90} : (2,1) \mapsto (-1,2)$
$\quad (0,0) \mapsto (0,0).$

Remark: Linear transformations always transform the origin to itself, i.e., $(0,0) \mapsto (0,0)$.

This transformation is illustrated in Fig. 24.2a.

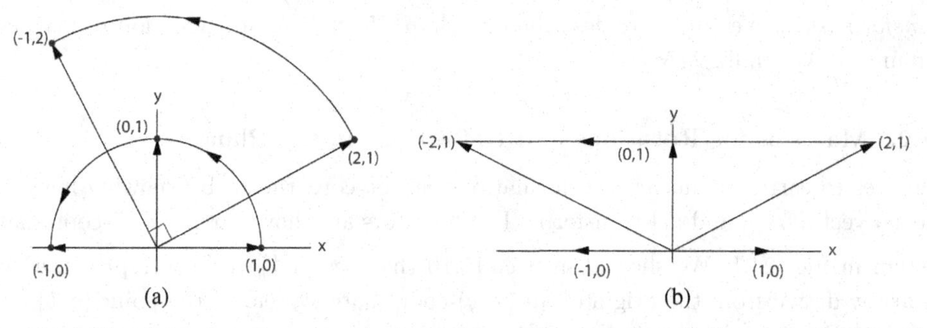

Fig. 24.2. Transformation of a vector determined by action on vectors (1,0) and (0,1): (a) 90 deg. counterclockwise rotation; (b) reflection in the y-axis.

b. Reflection in the y-axis, i.e., R_y:

$$(1,0) \rightarrow (-1,0)$$
$$(0,1) \rightarrow (0,1).$$

Therefore,

$$R_y = \begin{bmatrix} -1 & 0 \\ 0 & 1 \end{bmatrix}. \tag{24.2}$$

The point (2, 1) transforms as follows:

$$\begin{bmatrix} -1 & 0 \\ 0 & 1 \end{bmatrix} \begin{bmatrix} 2 \\ 1 \end{bmatrix} = \begin{bmatrix} -2 \\ 1 \end{bmatrix}$$

$$R_y : (2,1) \mapsto (-2,1)$$

i.e., this transformation is illustrated in Fig. 24.2b.
c. Reflection in a mirror at 45 deg. to the x-axis, i.e., R_{45}:

$$(1,0) \to (0,1)$$
$$(0,1) \to (0,1).$$

Therefore,

$$R_{45} = \begin{bmatrix} 0 & 1 \\ 1 & 0 \end{bmatrix}. \tag{24.3}$$

d. Now consider a compound transformation in which one first rotates by 90 deg. and then reflects in the y-axis, i.e., R_yS_{90}. Again we check to see where the two special vectors transform, i.e.,

$$(1,0) \to (0,1)$$
$$(0,1) \to (1,0).$$

Do you see that?

Therefore,

$$R_yS_{90} = \begin{bmatrix} 0 & 1 \\ 1 & 0 \end{bmatrix}. \tag{24.4}$$

Notice that this is the same matrix that we obtained in Matrix 24.3. Therefore, it must be that,

$$R_yS_{90} = R_{45}. \tag{24.5}$$

Problem:

1. Use the ideas in Chaps 19 and 21 to prove this result. Hint: Write S_{90} as the product of two reflections whose mirror lines intersect at 45 deg.

We can obtain the result of Eq. (24.4) in another way. Write the product of transformations as a product of matrices, i.e.,

$$R_yS_{90} = \begin{bmatrix} -1 & 0 \\ 0 & 1 \end{bmatrix} \begin{bmatrix} 0 & -1 \\ 1 & 0 \end{bmatrix} = \begin{bmatrix} 0 & 1 \\ 1 & 0 \end{bmatrix}.$$

The operation of matrix multiplication to obtain this result is described in Appendix 24A.

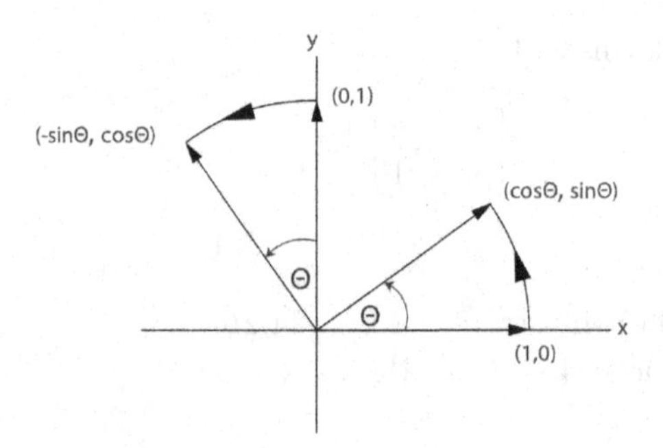

Fig. 24.3. Rotation by angle θ.

e. Rotation through angle θ (see Fig. 24.3)

$$(1,0) \to (\cos\theta, \sin\theta)$$
$$(0,1) \to (-\sin\theta, \cos\theta).$$

Therefore,

$$S_\theta = \begin{bmatrix} \cos\theta & -\sin\theta \\ \sin\theta & \cos\theta \end{bmatrix}. \tag{24.6}$$

Remark: If θ = 90 deg., Matrix 24.6 reduces to Matrix 24.1.

Problems:

For each of the following problems determine the matrix that carries out the given compound transformation. Check your result by multiplying the matrices for the elementary matrices that make up the compound transformation.

2. Rotate 90 deg. counterclockwise about the origin and then reflect in the x-axis, i.e., $R_x S_{90}$.
3. Reflect in the negative-leaning 45-deg. line ($y = -x$) and then rotate 180 deg., i.e., $S_{180} R_{-45}$.
4. Reflect in the y-axis followed by a 90 deg. clockwise rotation, i.e., $S_{-90} R_y$.
5. Rotate 45 deg. counterclockwise and then reflect in the 45-deg line, i.e., $S_{45} R_{45}$.

24.3. Transformations of Mr. Flatlands

A portrait of Mr. Flatlands is rendered in Fig. 24.4. His three vertices are: (0, 0), (1, 0) and (1, 1) and his eye is located at $(\frac{1}{2}, \frac{1}{2})$. We wish to see how Mr. Flatlands transforms under various isometries.

For example, let us subject him to S_{90} and R_y. To do this, we represent each of the vertices by a position vector of the same name, and then transform each of the position vectors, $\vec{v}$, by multiplying it by the matrix corresponding to the isometry.

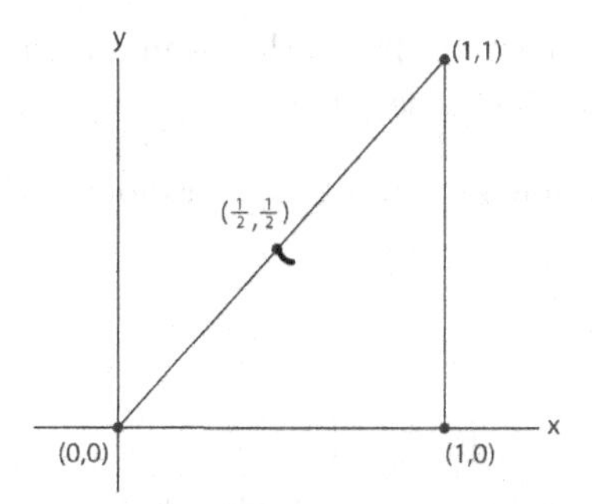

Fig. 24.4. Mr. Flatlands.

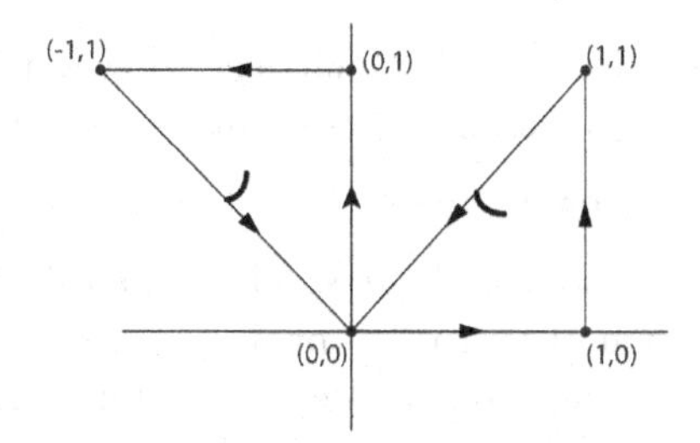

Fig. 24.5. Rotation of Mr. Flatlands through 90 deg. counterclockwise.

For a 90-deg. rotation,

$$S_{90}\vec{v} = \begin{bmatrix} 0 & -1 \\ 1 & 0 \end{bmatrix} \begin{bmatrix} 0 \\ 0 \end{bmatrix}, \begin{bmatrix} 1 \\ 0 \end{bmatrix}, \begin{bmatrix} 1 \\ 1 \end{bmatrix}, \begin{bmatrix} 1/2 \\ 1/2 \end{bmatrix} = \begin{bmatrix} 0 \\ 0 \end{bmatrix}, \begin{bmatrix} 0 \\ 1 \end{bmatrix}, \begin{bmatrix} -1 \\ 1 \end{bmatrix}, \begin{bmatrix} -1/2 \\ 1/2 \end{bmatrix}$$

The result of this transformation is shown graphically in Fig. 24.5. Notice that as one travels around the perimeter of Mr. Flatlands in a counterclockwise direction, the sequence of his image points are also counterclockwise.

b. Now consider a reflection in the y-axis.

$$R_y\vec{v} = \begin{bmatrix} -1 & 0 \\ 0 & 1 \end{bmatrix} \begin{bmatrix} 0 \\ 0 \end{bmatrix} \begin{bmatrix} 1 \\ 0 \end{bmatrix} \begin{bmatrix} 1 \\ 1 \end{bmatrix} \begin{bmatrix} 1/2 \\ 1/2 \end{bmatrix} = \begin{bmatrix} 0 \\ 0 \end{bmatrix}, \begin{bmatrix} -1 \\ 0 \end{bmatrix}, \begin{bmatrix} -1 \\ 1 \end{bmatrix}, \begin{bmatrix} -1/2 \\ 1/2 \end{bmatrix}$$

The result is shown in Fig. 24.6.

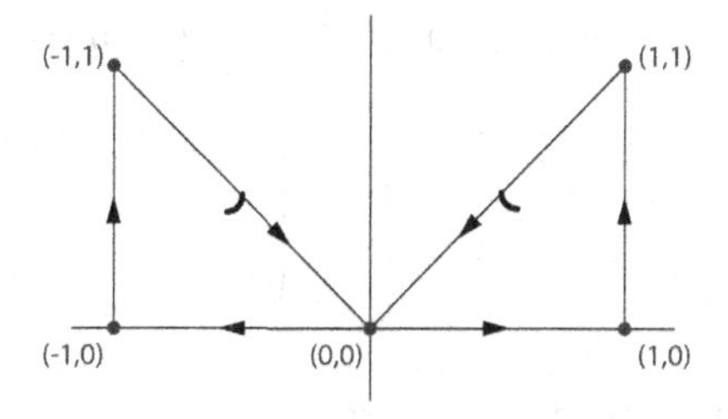

Fig. 24.6. Reflection in the y-axis of Mr. Flatlands.

This time as you travel around Mr. Flatlands in a counterclockwise direction, you travel around his image in a clockwise direction.

Remark: You will find, in general, that proper isometries preserve orientation while improper isometries reverse orientation.

Problem:

6. Transform Mr. Flatlands by a 45 deg. reflection in the negatively sloped line $y = -x$. Show that this transformation reverses orientation.

24.4. Homogeneous Coordinates

Translations can also be carried out with matrices. However, to do so requires us to redefine the way we represent points in the coordinate system. Instead of representing point (a, b) by vector $\begin{bmatrix} a \\ b \end{bmatrix}$ we now represent it by the triple $\begin{bmatrix} a \\ b \\ 1 \end{bmatrix}$ with a 1 in the third position, known as *homogeneous coordinates.* These coordinates have been designed to facilitate the representation of vectors in projective geometry, and we will explain the motivation for this representation in Sec. 24.8a. Furthermore, we also have the understanding that $\begin{bmatrix} a \\ b \\ 1 \end{bmatrix} = \begin{bmatrix} ka \\ kb \\ k \end{bmatrix}$ for any value of k. Therefore, if we encounter the vector $\begin{bmatrix} ka \\ kb \\ k \end{bmatrix}$ to find out what point in the coordinate system this vector pertains to, we must divide each element by the last element, i.e., k, so that a 1 again appears in the third position. The point (a, b) is then read off from the first two positions. For example,

$$\begin{bmatrix} 6 \\ 4 \\ 2 \end{bmatrix} = \begin{bmatrix} \frac{3}{2} \\ 1 \\ \frac{1}{2} \end{bmatrix} = \begin{bmatrix} 3 \\ 2 \\ 1 \end{bmatrix},$$

from which it follows that the point is $(3, 2)$.

24.5. Translations

Now a 3×3 matrix can be constructed to carry out translations. First consider the general 3×3 matrix,

$$\begin{bmatrix} a & b & c \\ d & e & f \\ g & h & i \end{bmatrix}. \tag{24.7}$$

The 2×2 sub-matrix defined by a, b, e, d carries out rotations and reflections as we did in Secs 24.2; c and f are translators; while i scales figures. The remaining two positions,

g and h, pertain to projections. The matrix,

$$E = \begin{bmatrix} 1 & 0 & 0 \\ 0 & 1 & 0 \\ 0 & 0 & 1 \end{bmatrix}$$

represents the identity, E, the isometry that leaves points where they are.

Now consider the matrix,

$$T_{3,2} = \begin{bmatrix} 1 & 0 & 3 \\ 0 & 1 & 2 \\ 0 & 0 & 1 \end{bmatrix}. \tag{24.8}$$

Since the 2×2 submatrix is the identity there are no rotations and reflections. Since $i = 1$ there is no scaling. Since $g = h = 0$ there are no projections. Since $c = 3$ and $f = 2$, this matrix translates a point 3 units to the right and 2 units up. To illustrate, use Eqn. (24.8) to transform Mr. Flatlands represented by the vectors,

$$\begin{bmatrix} 0 \\ 0 \\ 1 \end{bmatrix}, \begin{bmatrix} 1 \\ 0 \\ 1 \end{bmatrix}, \begin{bmatrix} 1 \\ 1 \\ 1 \end{bmatrix} \begin{bmatrix} \frac{1}{2} \\ \frac{1}{2} \\ 1 \end{bmatrix}.$$

Carrying out matrix multiplications in the usual way, yields the transformed vectors,

$$\begin{bmatrix} 0 \\ 0 \\ 1 \end{bmatrix}, \begin{bmatrix} 1 \\ 0 \\ 1 \end{bmatrix}, \begin{bmatrix} 1 \\ 1 \\ 1 \end{bmatrix}, \begin{bmatrix} \frac{1}{2} \\ \frac{1}{2} \\ 1 \end{bmatrix} \to \begin{bmatrix} 3 \\ 2 \\ 1 \end{bmatrix}, \begin{bmatrix} 4 \\ 2 \\ 1 \end{bmatrix}, \begin{bmatrix} 4 \\ 3 \\ 1 \end{bmatrix}, \begin{bmatrix} \frac{7}{2} \\ \frac{5}{2} \\ 1 \end{bmatrix}.$$

Make sure that you can carry out these matrix multiplications. Notice in Fig. 24.7 that Mr. Flatlands has been translated 3 units to the right and 2 units up without rotation or reflection.

Remark: Notice that the origin, $(0, 0)$ has transformed to $(3, 2)$ so this transformation must not be a linear transformation.

Placing the 2×2 matrix for a 90-deg. rotation into the matrix for $T_{3,2}$ results in the compound transformation: rotate 90 deg. counterclockwise and then translate, i.e,

$$\begin{bmatrix} 0 & -1 & 3 \\ 1 & 0 & 2 \\ 0 & 0 & 1 \end{bmatrix} = \begin{bmatrix} 1 & 0 & 3 \\ 0 & 1 & 2 \\ 0 & 0 & 1 \end{bmatrix} \begin{bmatrix} 0 & -1 & 0 \\ 1 & 0 & 0 \\ 0 & 0 & 1 \end{bmatrix} = T_{3,2}S_{90}.$$

Remark: In general, when placing the 2×2 matrix within the 3×3 matrix of homogeneous coordinates, the rotation or reflection represented by the 2×2 matrix is carried out first followed by the translation.

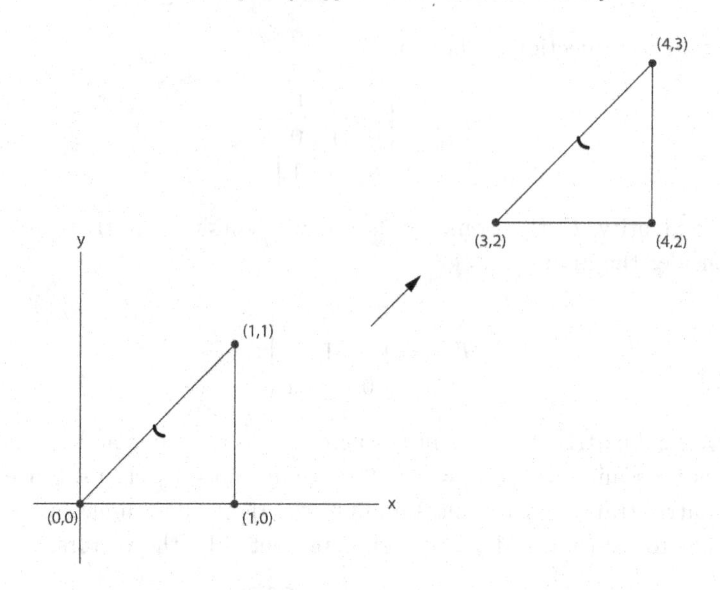

Fig. 24.7. Translation of Mr. flatlands by vector (3,2).

Problems:

Express the vertices of Mr. Flatlands in homogeneous coordinates, carry out the following transformations on him, and draw his new position in (x,y)-coordinates:

7. Rotate Mr. Flatlands by 90 deg. counterclockwise and then translate him 2 units to the right and 1 unit up.
8. Reflect Mr. Flatlands in the y-axis and then translate him 2 units to the left and 1 unit up.

24.6. Scaling

As mentioned before, the value of i determines the scaling. Let us see what the transformation,

$$Sc = \begin{bmatrix} 1 & 0 & 0 \\ 0 & 1 & 0 \\ 0 & 0 & 2 \end{bmatrix} \tag{24.9}$$

does to the unit square (see Fig. 24.8). Multiplying the vertex points of the square by Matrix 24.8 yields,

$$\begin{bmatrix} 0 \\ 0 \\ 1 \end{bmatrix}, \begin{bmatrix} 1 \\ 0 \\ 1 \end{bmatrix}, \begin{bmatrix} 1 \\ 1 \\ 1 \end{bmatrix}, \begin{bmatrix} 0 \\ 1 \\ 1 \end{bmatrix} \mapsto \begin{bmatrix} 0 \\ 0 \\ 2 \end{bmatrix}, \begin{bmatrix} 1 \\ 0 \\ 2 \end{bmatrix}, \begin{bmatrix} 1 \\ 1 \\ 2 \end{bmatrix}, \begin{bmatrix} 0 \\ 1 \\ 2 \end{bmatrix} = \begin{bmatrix} 0 \\ 0 \\ 1 \end{bmatrix}, \begin{bmatrix} \frac{1}{2} \\ 0 \\ 1 \end{bmatrix}, \begin{bmatrix} \frac{1}{2} \\ \frac{1}{2} \\ 1 \end{bmatrix}, \begin{bmatrix} 0 \\ \frac{1}{2} \\ 1 \end{bmatrix}.$$

Notice that the matrix scales the original square by one-half or $Sc_{(1/2)}$.

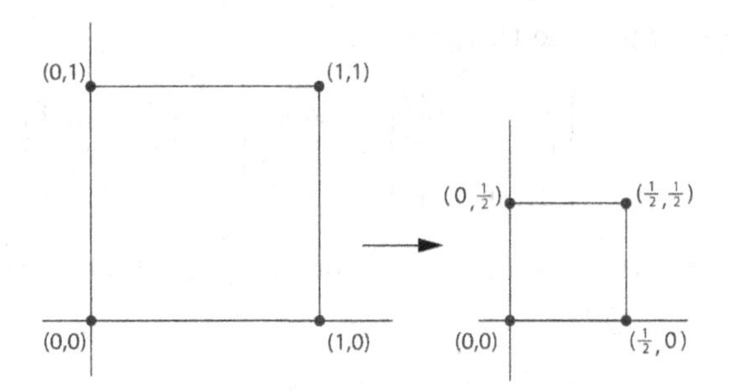

Fig. 24.8. Scaling a figure by one-half.

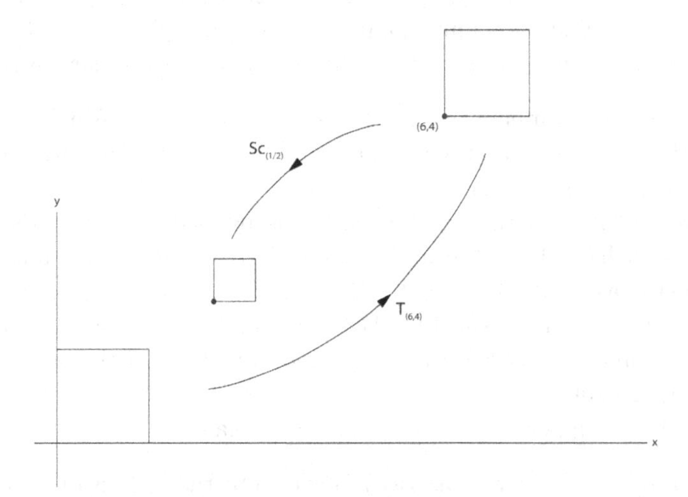

Fig. 24.9. The combination of translating a scaled copy of some figure.

Remark: Since '2' appears in the 3rd position of the transformed vectors, we must divide each element by 2 in order to identify the new position.

Remark: Each vertex point of the original square has been transformed to a point half of the distance to the origin.

Remark: We have seen that if $i = 2$, the figure scales by $\frac{1}{2}$. In general, if $i = k$, the figure scales by $\frac{1}{k}$.

How would you translate the unit square 3 units to the right and 2 units up, and then scale it by $\frac{1}{2}$ as shown in Fig. 24.9?

Here you have to be a little clever. Since you want the origin to translate to $(3, 2)$, you will first have to translate the origin of the square to $(6, 4)$ so that when it gets scaled, the origin will move half the distance to the origin, ending up at $(3, 2)$. The matrix to do this is,

$$\begin{bmatrix} 1 & 0 & 6 \\ 0 & 1 & 4 \\ 0 & 0 & 2 \end{bmatrix}.$$

This matrix can be shown to be the product of

$$Sc_{\frac{1}{2}}T_{6,4} = \begin{bmatrix} 1 & 0 & 0 \\ 0 & 1 & 0 \\ 0 & 0 & 2 \end{bmatrix} \begin{bmatrix} 1 & 0 & 6 \\ 0 & 1 & 4 \\ 0 & 0 & 1 \end{bmatrix} = \begin{bmatrix} 1 & 0 & 6 \\ 0 & 1 & 4 \\ 0 & 0 & 2 \end{bmatrix}.$$

Remark: Whenever translating a scaled figure, it is important to see where the origin of the original figure transforms.

24.7. Fractals

Let us now apply what we have learned about matrices to the Iterative Function System (IFS) of Chap. 23. Begin with a unit square with the letter 'p' within it. Divide the unit square into four quadrants and transform the contents of this square in three ways.

a. The square and its contents in Fig. 24.10a is scaled down to half of its size and placed in the upper left-hand quadrant of the original square, without rotation or reflection as shown in Fig. 24.10b.
b. The square and its contents in Fig. 24.10a is rotated 90 deg. counterclockwise, scaled down to half of its size and translated to the lower left quadrant of the original square, as shown in Fig. 24.10b.
c. The square and its contents in Fig. 24.10a is rotated by 180 deg., scaled down to half of its size and translated to the lower right quadrant of the original square, as shown in Fig. 24.10b.
d. The square in the upper right quadrant remains blank.

We can now easily construct the matrices to carry out these transformations.

For (a): After translating the square and its contents to the upper left quadrant without rotation or reflection, and scaling by $\frac{1}{2}$, the origin is located at coordinate $(0, \frac{1}{2})$. But since we need to scale by $\frac{1}{2}$, the origin should be translated first to $(0, 1)$ so that when scaling brings it half the distance to the origin, it will be in the correct place. This sequence of operations yields the result in the upper left square of Fig. 24.10b as

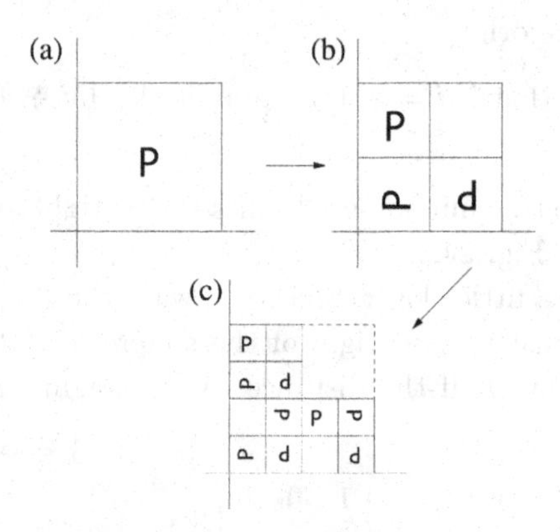

Fig. 24.10. Steps in creating a fractal from the Iterative Function System: (a) stage 0; (b) stage 1; (c) stage 2.

carried out by the matrix.

$$Sc_{\frac{1}{2}}T_{0,1} = \begin{bmatrix} 1 & 0 & 0 \\ 0 & 1 & 1 \\ 0 & 0 & 2 \end{bmatrix}. \tag{24.10}$$

For (b): We first rotate the unit square and its contents by 90 deg. counterclockwise, i.e., S_{90}, translate the origin to $(1, 0)$, and then scale by $\frac{1}{2}$. This ensures that after scaling, the origin is in the correct position, $(\frac{1}{2}, 0)$. This sequence of operations yields the result in the lower left quadrant of Fig. 24.10b) as carried out by the matrix

$$Sc_{\frac{1}{2}}T_{1,0}S_{90} = \begin{bmatrix} 0 & -1 & 1 \\ 1 & 0 & 0 \\ 0 & 0 & 2 \end{bmatrix}. \tag{24.11}$$

For (c): We first rotate the unit square and its contents by 180 deg., then translate so that the origin moves to $(2, 1)$, followed by a scaling by $\frac{1}{2}$. This ensures that the origin moves to its correct position $(1, \frac{1}{2})$. This sequence of operations yields the result in the lower right square of Fig. 24.10b as carried out by the matrix

$$Sc_{\frac{1}{2}}T_{2,1}S_{180} = \begin{bmatrix} -1 & 0 & 2 \\ 0 & -1 & 1 \\ 0 & 0 & 2 \end{bmatrix}. \tag{24.12}$$

We have now generated the picture shown in Fig. 24.10b. If we apply Matrix 24.9 to this figure, it scales again by $\frac{1}{2}$ and moves it to the upper left quadrant of the original unit square in Fig. 24.10c. Applying Matrix 24.11 to Fig. 24.10b rotates the figure 90 deg., scales it and moves it to the lower left quadrant of the original unit square as shown in Fig. 24.10c. Matrix 24.12 rotates Fig. 24.10b by 180 deg., scales it by $\frac{1}{2}$ and moves it to the lower right quadrant of the original square in Fig. 24.10c. The result of this stage of the transformation is shown in Fig. 24.10c. Now we can apply these three matrices over and over again to obtain an advanced stage of the fractal. At the end of the day, the fractal will appear as in Fig. 24.11 which was illustrated in Fig. 23.7a.

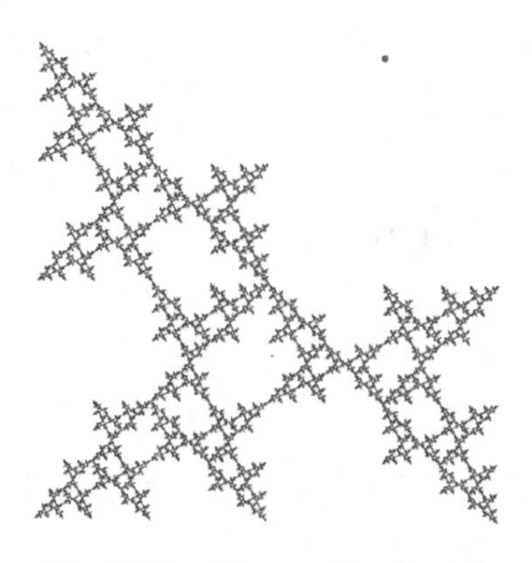

Fig. 24.11. The resulting fractal.

Remark: The final result will be independent of the contents of the unit square that you have started with. So instead of a 'p', you could have used a scribble or even a black square as in Sec. 23.6.

Remark: Seven or eight stages should be enough to articulate the fractal.

Problem:

9. Find the three matrices from the IFS system that resulted in the fractal of Fig. 23.6, i.e.,

a) Translate the unit square with a 'p' and scale it by $\frac{1}{2}$ so that it fits into the bottom right square.

b) Rotate the initial unit square with 'p' by 90 deg., translate, and scale it by $\frac{1}{2}$ so that it fits in the upper left square.

c) Rotate the initial unit square with 'p' by 180 deg., translate, and scale it by $\frac{1}{2}$ so that it fits in the lower left square.

24.8. Projections

24.8.1 Homogeneous coordinates revisited

To understand how matrices can represent projections, we must first return to the meaning of homogenous coordinates. Consider a line from the origin to infinity in the direction of vector $(2, 1)$ as shown in Fig. 24.12. As the points on the line go to infinity, the position vector progresses in steps: $(2, 1), (4, 2), (8, 4), (16, 8), \ldots, (2k, k)$. These points are represented in homogeneous coordinates by,

$$(2, 1, 1), (4, 2, 1), (8, 4, 1), (16, 8, 1), \ldots, (2k, k, 1).$$

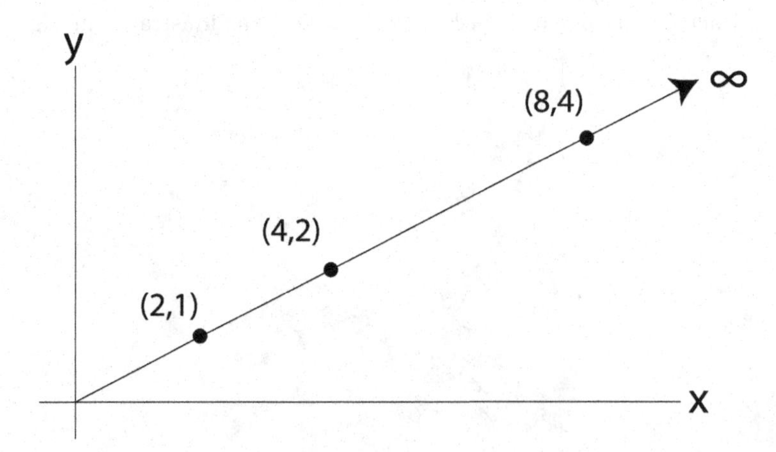

Fig. 24.12. Homogeneous coordinates define the point at infinity in any given direction.

Alternatively, we can divide these coordinates by the second coordinate to get,

$$(2,1,1), \left(2,1,\frac{1}{2}\right), \left(2,1,\frac{1}{4}\right), \left(2,1,\frac{1}{16}\right), \ldots, \left(2,1,\frac{1}{k}\right).$$

Now you see what happens if $k \to \infty$. The coordinates approach, $(2,1,0)$ in the sense of a limit. So we have achieved the major objective of projective geometry, that is to make inifinity tangible and operational. We now see that any point on the line at infinity in projective geometry in the direction of the vector (a,b) can be represented in homogeneous coordinates by $(a,b,0)$.

24.8.2 Transformation of points on the line at infinity

Now consider the 3×3 matrix (24.7) written as the identity except for $g = 1$ and $h = \frac{1}{2}$, i.e.,

$$\begin{bmatrix} 1 & 0 & 0 \\ 0 & 1 & 0 \\ 1 & \frac{1}{2} & 1 \end{bmatrix}. \tag{24.13}$$

Let us see what this matrix does to three randomly selected points on the line at infinity in the directions of: $(1,2)$, $(0,1)$, $(1,1)$. Writing them in homogeneous coordinates as $(1,2,0)$, $(0,1,0)$, $(1,1,0)$ and transforming them by Matrix (20.13) yields,

$$(1,2,2), \left(0,1,\frac{1}{2}\right), \left(1,1,\frac{3}{2}\right).$$

Dividing these by their third coordinate gives the result,

$$\left(\frac{1}{2},1,1\right), (0,2,1), \left(\frac{2}{3},\frac{2}{3},1\right)$$

which corresponds to the three points, : $(\frac{1}{2},1)$, $(0,2)$, $(\frac{2}{3},\frac{2}{3})$.

Remark: Notice that whereas the original three points were infinite, these transformed points are finite.

Remark: Even more significant, the three transformed points all lie on the same line. You can see this by finding the equation of the line joining the first two points and then show that point 3 lies on that line. We use the point-slope method where

$$m = \frac{2-1}{0-\frac{1}{2}} = -2$$

and the point is $(0,2)$. The line is then,

$$y - 2 = -2(x - 0) \quad \text{or} \quad y = -2x + 2.$$

Replacing the third point into this equation,

$$\frac{2}{3} = -\frac{4}{3} + 2 \text{ which checks.}$$

This line must then be the *horizon line*. By choosing different values of g and h in Matrix 24.9, we can obtain all possible horizon lines.

Problem:

10. For Matrix 24.7 written as the identity except for $g = 1$ and $h = -1$, i.e.,

$$\begin{bmatrix} 1 & 0 & 0 \\ 0 & 1 & 0 \\ 1 & -1 & 1 \end{bmatrix}.$$

Choose three points on the line at infinity, $(1, 2)$, $(0, 1)$, $(1, 1)$, and show that they transform to three finite points on a line. Find the equation of the line.

24.8.3 Railroad tracks receding to infinity

Now let us see what happens to a pair of railroad tracks as they recede to infinity. Consider the two parallel lines, L_1, L_2, in Fig. 24.13a.

The coordinates of L_1 and L_2 can be expressed as $L_1 : (0, t)$, L_2, can be expressed as $L_2 : (1, t)$ for $t > 0$. We express the coordinates of these lines in homogeneous coordinates as: $L_1 : (0, t, 1)$, $L_2 : (1, t, 1)$ and transform them by the Projective Matrix 24.12.

The result is: $L_1 : (0, t, 1 + \frac{t}{2})$, and $L_2 : (1, t, 1 + \frac{t}{2})$.

Dividing by the third coordinate,

$$L_1 : \left(0, \frac{t}{1 + \frac{t}{2}}\right), \quad L_2 : \left(\frac{1}{1 + \frac{t}{2}}, \frac{t}{1 + \frac{t}{2}}\right).$$

Corresponding to $t = 0$ are the points L_1: $(0, 0)$ and L_2: $(1, 0)$.

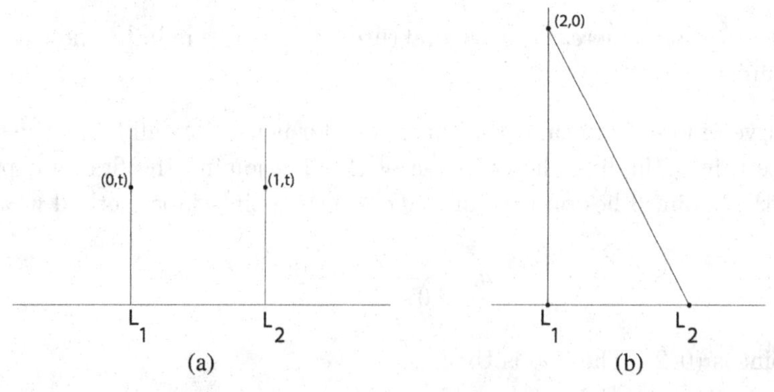

Fig. 24.13. (a) Parallel railroad tracks receding to infinity; (b) the railroad meet at a vanishing point on the horizon line.

Corresponding to large values of t, in the sense of a limit you can see that L_1 and L_2 both approach $(0,2)$ as $t \to \infty$. L_1 is clearly the line segment $x = 0$, $0 < y < 2$. What about L_2? To show that it is a line segment, let

$$x = \frac{1}{1+\frac{t}{2}}, \quad y = \frac{t}{1+\frac{t}{2}}.$$

Solving for t in terms of x, after some algebra,

$$t = \frac{2}{x} - 2, \quad 0 \leq t < \infty.$$

Replacing this in the equation for y, after some algebra,

$$y = 2 - x.$$

So we have found that the railroad tracks now appear to converge to $(0,2)$ in perspective as shown in Fig. 24.13b.

We conclude that the projective matrices are an excellent way to represent the subject of projective geometry.

Appendix 24A An Introduction to Matrices

A Matrix is a rectangular array of numbers. For example,

$$A = \begin{bmatrix} a_{11} & a_{12} \\ a_{21} & a_{22} \\ a_{31} & a_{32} \end{bmatrix}$$

is a 3×2 matrix since it has 3 rows and 2 columns. The symbol a_{ij} stands for the real number in the ith row and j^{th} column. In our study we shall have need only for square matrices, i.e., matrices with the same number of rows as columns. We have seen in this chapter that square matrices can be used to represent isometries. Just as we have seen that transformations can by multiplied, matrices can also be multiplied. Consider the product $AB = C$ of the two 2×2 matrices,

$$AB = \begin{bmatrix} -1 & 0 \\ 0 & 1 \end{bmatrix} \begin{bmatrix} 0 & -1 \\ 1 & 0 \end{bmatrix} = \begin{bmatrix} -1 \times 0 + 0 \times 1 & (-1) \times (-1) + 0 \times 0 \\ 0 \times 0 + 1 \times 1 & 0 \times (-1) + 1 \times 0 \end{bmatrix} = \begin{bmatrix} 0 & 1 \\ 1 & 0 \end{bmatrix} = C.$$

According to Sec. 24.2, B is the matrix of the transformation S_{90} while A is the matrix representation of R_y, in which case C represents the matrix of the reflection in the 45 deg. line, R_{45}. We see that the element in the 1st row and 1st column of the product matrix, c_{11}, is the product and sum of the corresponding elements of row 1 of A and column 1 of B. The element in the 2nd row, 1st column of C, c_{21}, is the product and sum of the corresponding elements of row 2 of A and column 1 of B. In general, the element in the ith row and j^{th} column of the product matrix, c_{ij}, is the sum and product of corresponding elements of the ith row of A and the jth column of B.

Matrices with a single column are used to represent vectors. For example, the position vector $\bar{v} = (2,1)$ or point $(2,1)$ can be represented by the matrix $\left[\begin{smallmatrix} 2 \\ 1 \end{smallmatrix}\right]$. Matrices can be used to transform vectors through the operation of multiplication of matrices. For example, if you wish to use the matrix $\left[\begin{smallmatrix} 0 & -1 \\ 1 & 0 \end{smallmatrix}\right]$ to rotate vector $\left[\begin{smallmatrix} 2 \\ 1 \end{smallmatrix}\right]$ by 90 deg. counterclockwise about the origin of an (x, y)-coordinate system, simply multiply the matrix for 90 deg. rotations by the vector using the above procedure as follows:

$$\begin{bmatrix} 0 & -1 \\ 1 & 0 \end{bmatrix} \begin{bmatrix} 2 \\ 1 \end{bmatrix} = \begin{bmatrix} 2 \times 0 + 1 \times (-1) \\ 2 \times 1 + 1 \times 0 \end{bmatrix} = \begin{bmatrix} -1 \\ 2 \end{bmatrix}$$

and $(-1, 2)$ is the result.

APPENDIX A

PROJECTIVE GEOMETRY CONSTRUCTIONS

A1. Introduction: Projective Transformations

Projective Geometry is the mathematical foundation of the theory of perspective. It is of great importance to architects and designers on the one hand and to the foundations of mathematics on the other. In fact, projective geometry had its origin in the work of artists and architects [EdwL], [Whi]. Figure A1.1 shows the projective transformation of a series of points A, B, C, D on line x to points A', B', C', D' on line x' with the point of projection being O. In Fig. A1.2 O is where the eye of an artist is located as she renders a scene on the horizontal plane by projecting it onto the artist's canvas. The line at infinity in the horizontal plane is projected to the horizon line h on canvas, and parallel lines receding to infinity meet on the horizon line in the plane of the painting. Figure A1.3 shows the rendering of a three-dimensional scene. In this

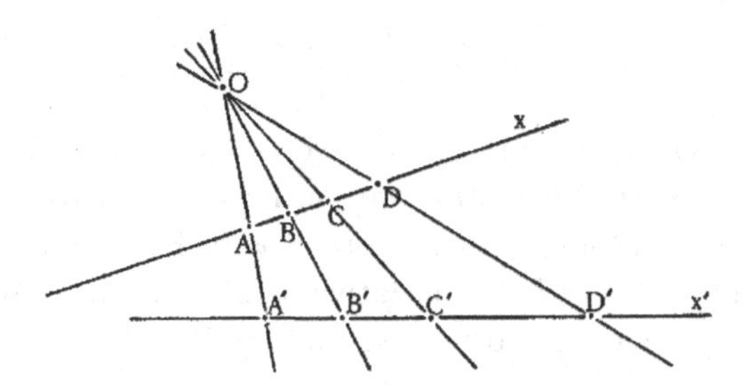

Fig. A1.1. Projection of line x to line x' through O.

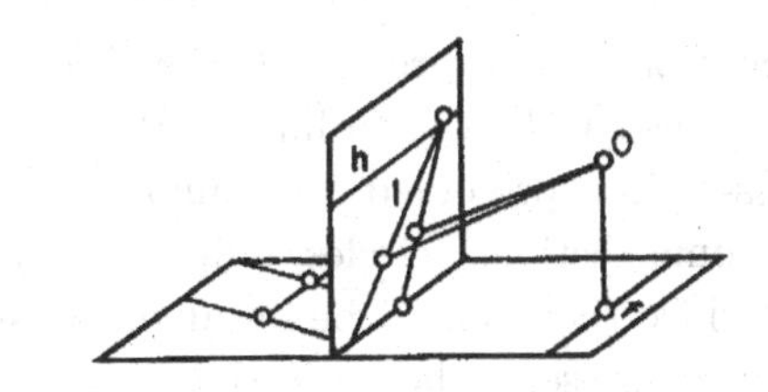

Fig. A1.2. A road receding to infinity depicted as converging to a point on the horizon line.

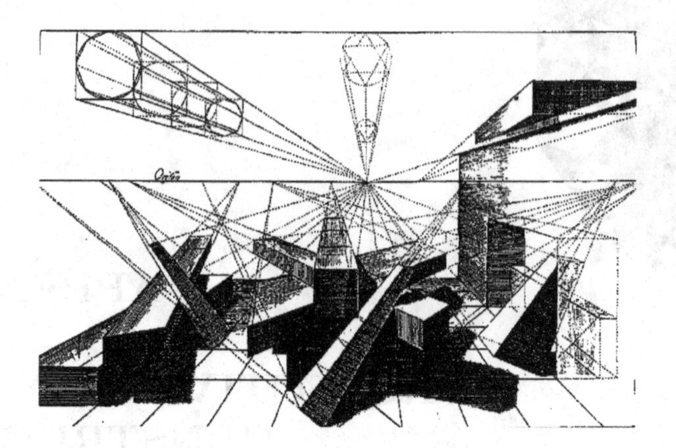

Fig. A1.3. Projective rendering of a 3D scene.

picture, all pairs of parallel lines meet on the horizon line. In fact, it is the ability of projective geometry to represent infinity in finite terms that is so valuable.

In Euclidean geometry of the plane, two figures are congruent (equivalent) if you can transform one by some combination of a movement in the plane and a reflection and superimpose the result of the transformation onto the other figure so that they match point for point. In Euclidean geometry, certain properties of the original figure such as the angle between lines and edge lengths remain unchanged (invariant) under this transformation. In projective geometry, two figures are considered to be equivalent if you can obtain one from the other by a sequence of projective transformations through points $O, O', O'' \ldots$ similar to the transformations on the artist's canvas. As a result of these transformations, angles and lengths generally change although there is one geometrical quantity that does not change called the *cross-ratio* λ. In Fig. A1.1, the cross-ratio is defined as,

$$\lambda = \frac{AB}{BC} \div \frac{AD}{DC}. \tag{A1.1}$$

This is easy to remember; in the first fraction we take a journey from A to C stopping at B to rest, and in the second fraction we journey from A to C, overshooting and stopping at D. If A, B, C, D are chosen in order as in Fig. A1.1, then AB, BC, and AD will be positive numbers with DC negative making λ a negative number. Since λ is preserved under projective transformations, in Fig. A1.1,

$$\lambda = \frac{AB}{BC} \div \frac{AD}{DC} = \frac{A'B'}{B'C'} \div \frac{A'D'}{D'C'}. \tag{A1.2}$$

Under projective transformations we are free to choose three points, e.g., A, B, C on line x and their transformations, A', B', C', on line x'. Points D and D' must then be positioned so that the cross-ratio is preserved; we cannot choose them arbitrarily.

It turns out that all conic sections (circles, ellipses, parabolas and hyperbolas) are projectively equivalent since they can all be obtained by slicing a double cone as shown in Fig. A1.4. Each slice represents the projection of the double cone onto a plane from point O at the vertex of the cone. Depending on the orientation of the plane, the

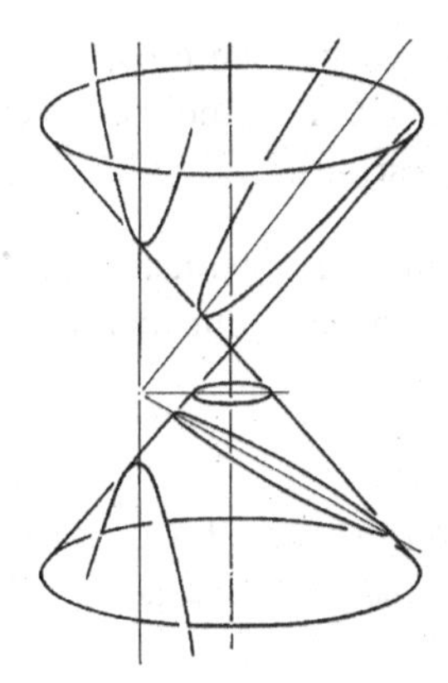

Fig. A1.4. Conic sections created from intersections of a cone by planes.

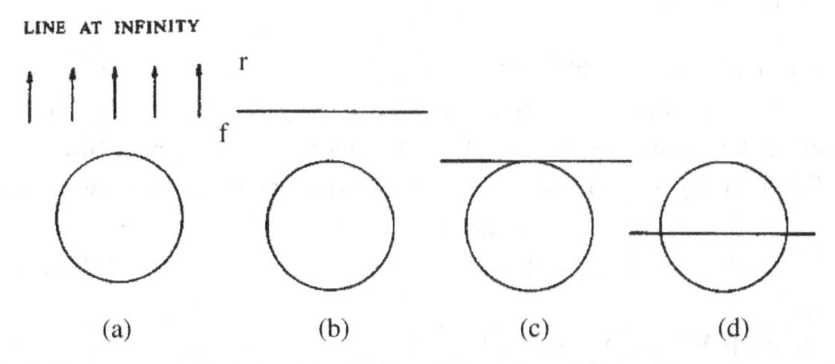

Fig. A1.5. A circle juxtaposed with a line in four different positions. The various conics are gotten by projecting the line to infinity.

projected image is either a circle, ellipse, parabola or hyperbola. Figure A1.5 shows four circles juxtaposed next to a line. Let the line be transformed by a projective transformation to the line at infinity. In Fig. A1.5a, the first circle remains unchanged since its line is already at infinity; the second circle, in Fig A1.5b, transforms to an ellipse; the third circle, in Fig. A1.5c, with the line tangent to the circle, opens up to a parabola; while the last circle, in Fig. A.15d, with the line intersecting the circle, breaks open to a hyperbola.

We will describe a series of constructions inspired by the excellent book *Projective Geometry* by Olive Whicher, using a number of figures from the book, without going deeply into the underlying theory. These constructions will show that space has certain intrinsic properties and constraints and is not completely freewheeling.

Let us begin with Construction A1.

A1.1. Construction 1: Cross-ratio

Choose two non-parallel lines x and x' as in Fig. 1.1 and point O of projection. Place four points A, B, C, D on x and find their images: A', B', C', D' with respect to O on x'. Use a ruler to measure the lengths AB, BC, AD, DC (it is more accurate to read the measurements in millimeters) and use Eq. A1.1 to compute the cross-ratio, λ. Then verify that the transformation preserves the cross-ratio by computing the cross-ratio of A', B', C', D' and showing that Eq. A1.2 holds.

Answer the following two questions:

(1) Where does the point at infinity on x transform to on x'?
(2) Which point on x transforms to the point at infinity on x'?

A2. Pappus' Theorem

Theorem (Pappus): *Given two lines and three points on each line, by connecting the points in the manner described below, a third line, the Pappus line, appears through the intersection points.*

We describe this theorem by using an analogy to the color wheel. Beginning with two arbitrary lines and two sets of three points (see Fig. A2.1), a third line mysteriously appears by the following construction:

a. Consider an arbitrary pair of lines labeled a, b. Place three arbitrary points labeled R_1, B_1, Y_1 on line a and three other arbitrary points labeled R_2, B_2, Y_2 on line b where R, B and Y can be thought of as the primary colors red, blue and yellow.
b. Connect R_1 with B_2 and B_1 with R_2. The intersection of these two lines is labeled P for purple (mixture of red and blue).
c. Connect R_1 with Y_2 and Y_1 with R_2. Label the intersection of these two lines, O for orange.
d. Connect B_1 with Y_2 and Y_1 with B_2. Label the intersection G for green.
e. The points P, O, G lie on a third line c which mysteriously appears.

Remark 1: Note that the sequence of six points, $R_1B_2Y_1R_2B_1Y_2R_1$, define six lines which form a six-sided closed figure or star-hexagon where P, O, G lie at the intersection points of opposite edges of this hexagon.

Remark 2: We are using upper case letters to represent points and lower case letters to represent lines.

Remark 3: The assignment of R, B, Y to the three points on each line is arbitrary. If the labeling were different the new intersections would still lie on a line although the line would be different.

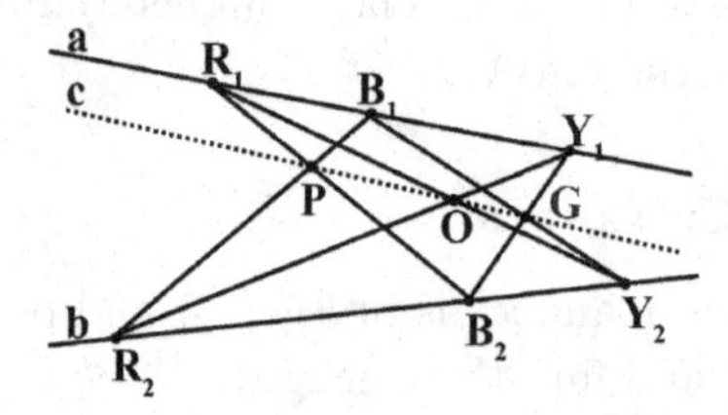

Fig. A2.1. Three points are placed on line a and line b. When they are interconnected a third line, the Pappus line, emerges.

Construction A2.1: Demonstrate Pappus' Theorem by choosing a pair of lines a, b and three points on each line and show that your construction leads to a third line, c. Identify the hexagon.

Remark 4: Depending on where you place the points and in which order you name them, the third line might be off the page so you may have to hunt for points which avoid this problem. The same holds for other projective constructions.

A3. The Dual of Pappus' Theorem

Theorem (Pappus' Dual): Begin with two points and two sets of three lines radiating from these two points. By generating three lines from the joins of the two sets of lines in a manner described below, a mysterious third point appears as shown in Fig. A3.1.

In projective geometry, for any construction or theorem in the plane there is a *dual* construction or theorem where the word 'line' is replaced by the word 'point' and 'point' is replaced by 'line'. Compare Pappus' Theorem in Construction A2. with its dual above.

To create the dual of Pappus' theorem:

a. Begin with an arbitrary pair of points labeled A and B.
b. Consider a pencil of three lines labeled r_1, b_1, y_1 incident to A and r_2, b_2, y_2 incident to B.
c. The intersections of r_1 with b_2, and b_1 with r_2 defines a line labeled p.
d. The intersections of r_1 with y_2, and y_1 with r_2 defines a second line labeled o.
e. The intersections of b_1 with y_2, and y_1 with b_2 defines a third line labeled g.
f. The three lines, p, o, g meet at a third point C which mysteriously appears.

Remark: Notice that the intersections of the six line pairs define six points which are the vertices of the shaded hexagon at the intersection points of the pairs of lines, $r_1 b_2 y_1 r_2 b_1 y_2 r_1$.

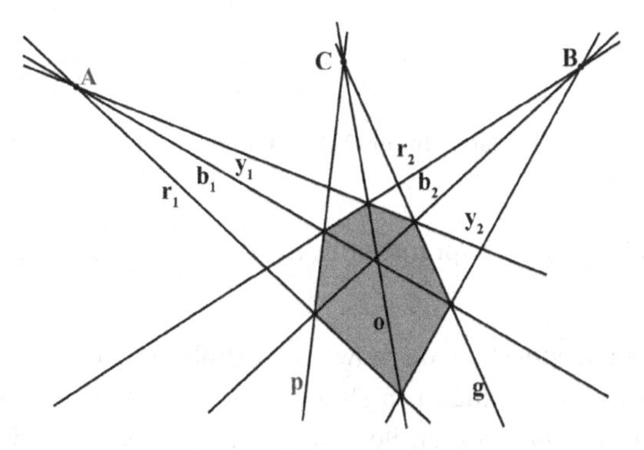

Fig. A3.1. Three lines are projected from two points A and B to find a third point C illustrating the dual to Pappus' Theorem.

Construction A3.1: Illustrate the dual of Pappus' Theorem by choosing two points A and B and three lines through these two points leading to a third point C. Identify the hexagon.

A4. Pascal's Theorem

In Fig. A4.1, two sets of three points, R_1, B_1, Y_1 and R_2, B_2, Y_2 are placed on an ellipse or a circle and Pappus' construction is carried out to show that once again P, O, G lie on a line, the so-called *Pascal line*, and a hexagon is again inscribed in the ellipse.

This construction goes by the name, *Pascal's Theorem.* You will notice that once a hexagon is inscribed in the ellipse or circle, opposite sides meet at the three points on the Pascal line. In Fig. A4.2, we show another illustration of Pascal's Theorem, but this time the three pair of opposite edges, labeled 1, 2, 3, meet at the three points on the Pascal line.

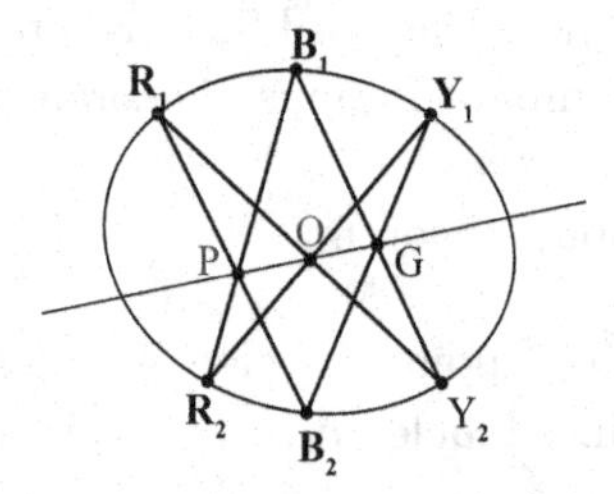

Fig. A4.1. Two sets of three points labeled by the primary colors are placed in an ellipse. The interconnections of these points create Pascal's line.

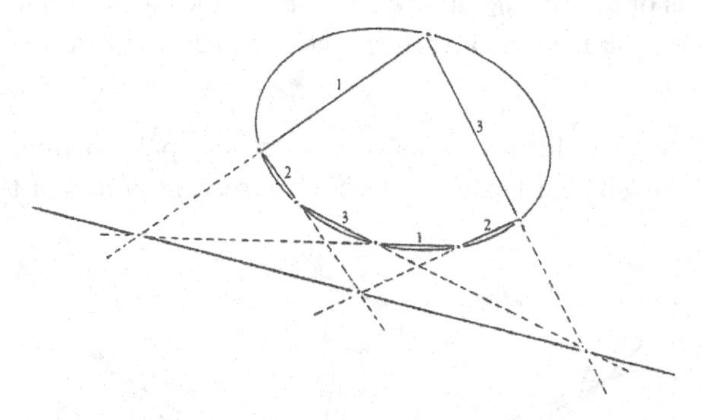

Fig. A4.2. A hexagon is placed in an ellipse. Opposite edges meet on Pascal's line.

Construction A4.1: Place six points anywhere on a circle or ellipse and draw the Pascal line.

Construction A4.2: Depending on how one numbers the six points on an ellipse, which are the vertices of a hexagon, a different Pascal line results. In Fig. A4.3, six equidistant points are placed on a circle, and the hexagons are illustrated that are the result of the twelve possible orderings of the points. Draw the Pascal line defined by the intersection of opposite edges for three of these hexagons.

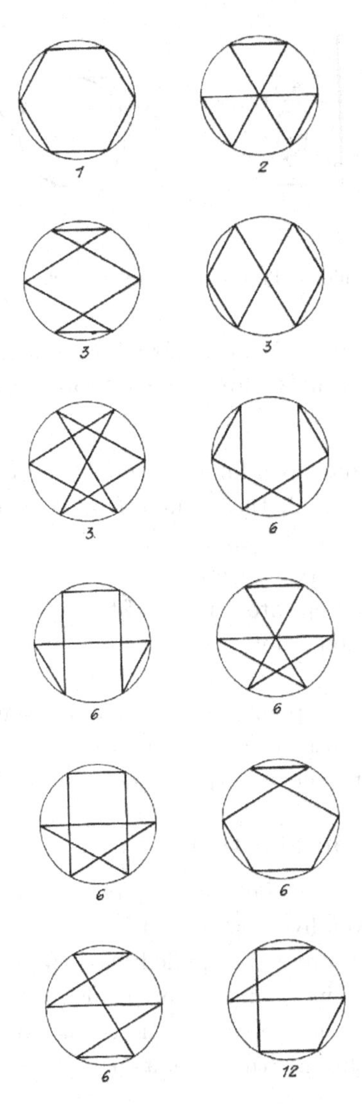

Fig. A4.3. Six points are equally distributed around a circle. A hexagon can be formed in 12 ways from these points.

A5. Branchion's Theorem

Branchion's Theorem is the dual to Pascal's Theorem.

In Pascal's Theorem, we saw that the points on an ellipse or circle are numbered $1, 2, 3, 4, 5, 6$ and the six points are connected in any order to define a hexagon. Opposite edges of the hexagon meet at three points on a line (the Pascal line) with the location of the line dependent on the order.

In *Branchion's Theorem*, the points on a conic are numbered 1 through 6 in any order, and six lines are drawn tangent to the six points and connected to form a hexagon. Opposite points on this circumscribing hexagon meet at a common point (the Branchion point) with the point dependent on the order.

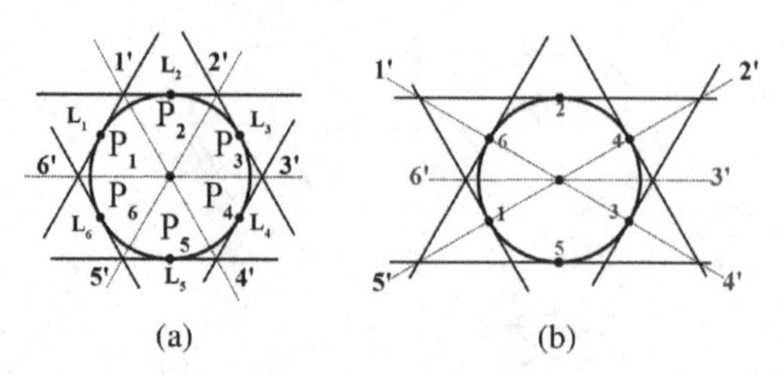

Fig. A5.1. Illustration of Branchion's Theorem for (a) consecutive points on around a circle; (b) non-consecutive points.

In Fig. A5.1, two circles are shown with six points approximately evenly distributed around the circumference of a circle. Branchion's theorem is illustrated for two different orderings of these points.

a. Tangent to each of the points is a line numbered the same as its tangent point.
b. Where line 1 intersects line 2, the point of intersection will be a vertex of the hexagon, and that vertex will be labeled $1'$.
c. Line 2 intersects line 3 at vertex $2'$, etc.
d. By Branchion's Theorem, opposite vertices of the hexagon, i.e., $1'$ and $4'$; $2'$ and $5'$; $3'$ and $6'$ meet at a common point, the Branchion point.

Construction A5.1: Place six points, approximately evenly spaced, around the circumference of a circle and construct a Branchion hexagon for two different orderings of the points, and show that opposite vertices meet at the Branchion point.

A6. Construction of a Conic Section

Pappus' or Pascal's Theorem can be used to draw the unique conic (ellipse, circle, hyperbola and parabola) given five points on it.

In Fig. A4.1, Pappus' Theorem was applied to an ellipse. In fact, Pappus' Theorem works for two sets of three points lying on any conic section. Using this theorem and starting with five arbitrary points on the plane as shown in Fig. A6.1, we construct the unique conic that goes through these points as follows:

a. Place five points arbitrarily in the plane and label them R_1, B_1, Y_1 and R_2, B_2 in any order. The problem is posed to find a sixth point on the conic labeled Y_2.

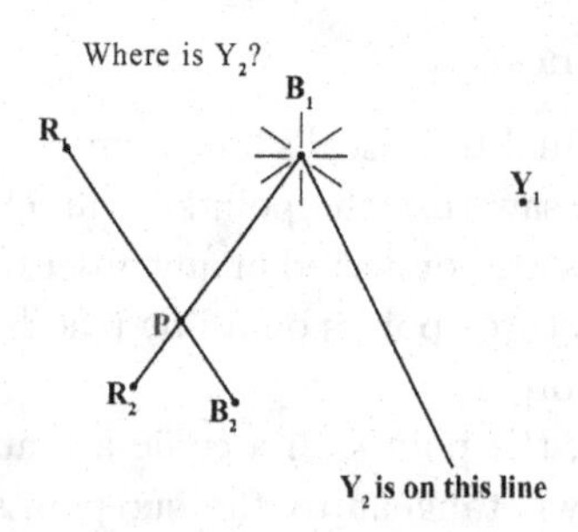

Fig. A6.1. Given five points in the plane, construct additional points on the unique conic through these points.

b. Connect R_1 with B_2 and B_1 with R_2. The intersection of these two lines is labeled P for purple.
c. Draw an arbitrary line from a pencil of lines through B_1. Somewhere on this line will be the missing point Y_2.
d. Connect Y_1 with B_2, and where it intersects the line through B_1 from Step c, label this point G.
e. According to the Pappus construction, P and G lie on the mysterious third line. Draw this line.
f. The line from R_2 to Y_1 intersects the line between P and G at O.
g. The line from R_1 to O intersects the line through B_1 in Step c at Y_2, and this is the missing point. It lies on the unique conic that goes through the five starting points.
h. Now repeat this construction for other lines from the pencil of lines through B_1 and you will be able construct all of the points on the conic.

Remark: Pappus' Theorem works for all conics. In fact, even the pair of lines in Construction A2 can be thought of as the extreme case of an hyperbola, where the hyperbola degenerates to its pair of asymptotes. Since all conics are projective transformations of each other, any theorem of projective geometry that works for one conic works for the others. For example, when Pappus' Theorem is applied to an ellipse or a circle it is called Pascal's Theorem.

Construction A6.1: Beginning with five arbitrary points, use this procedure to construct at least two additional points on the unique conic that goes through these points.

A7. Desargues Theorem

Desargues Theorem arises from the following construction:

a. Begin with a pair of triangles with vertices labeled ABC and $A'B'C'$ as in Fig. A7.1, with the constraint that lines AA', BB' and CC' meet at a common point O.
b. The three pairs of lines AB and $A'B'$; BC and $B'C'$; and CA and $C'A'$ intersect at three points.

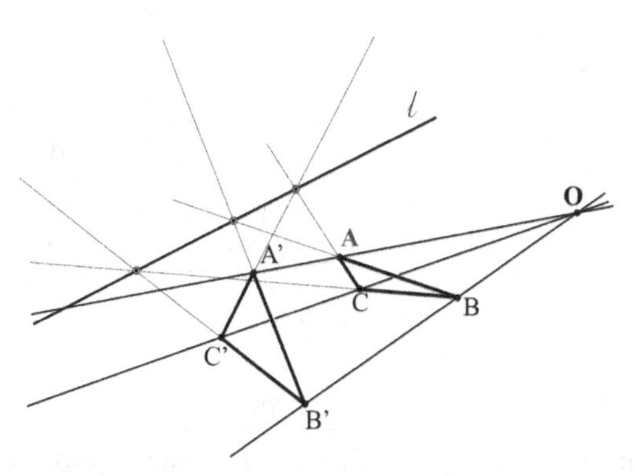

Fig. A7.1. Illustration of Desargues Theorem in which if two triangles are in perspective from point O, their corresponding edges meet at three points on a line l.

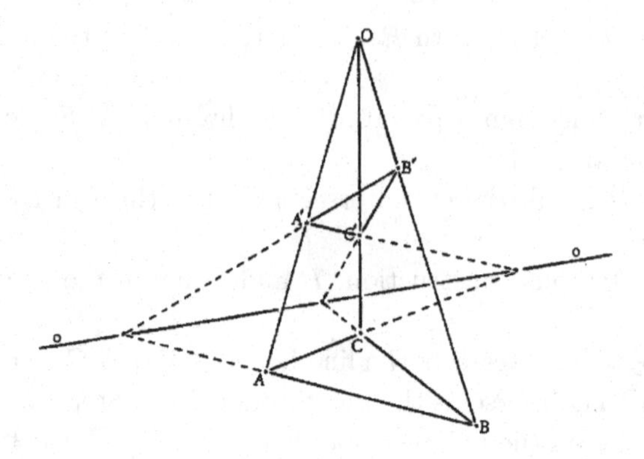

Fig. A7.2. Illustration of Desargues Theorem in 3-D.

c. These three points lie on a line, the Desargues line (they are collinear).

Remark: The reason why this theorem works can be seen by looking at a three-dimensional version of Desargues Theorem in Fig. A7.2. (Can you see it?)

Construction A7.1: Choose a pair of triangles whose corresponding vertices project to a common point O and generate the line l defined by Desargues Theorem.

A8. A Hexagonal Net

Beginning with four arbitrary lines we can create a hexagonal net of lines as follows:

a. Begin with four non-collinear but otherwise arbitrary lines (see Fig. A8.1a). One of these lines is designated as the horizon h.
b. A point is defined by the intersection of a pair of lines.
c. Lines can be drawn through already defined pairs of points.
d. In this way, the hexagon in Fig. A8.1b can be derived from Fig. A8.1a.
e. Continuing this process, you are able to generate a net of regular hexagons as in Fig. A8.2.

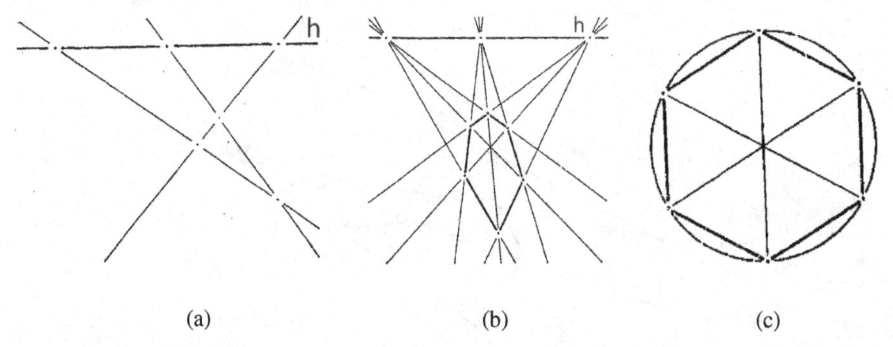

Fig. A8.1. (a) Four arbitrary lines are given with the horizontal line distinguished as the horizon line h; (b) the three points on the horizon line h are the vanishing points of pairs of opposite edges of a hexagon in perspective; (c) a regular hexagon.

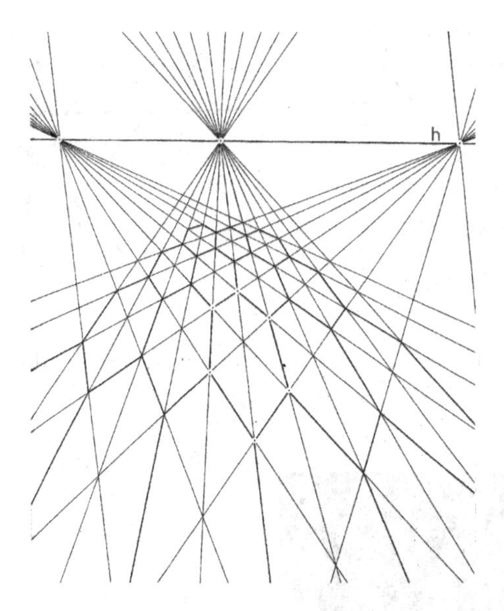

Fig. A8.2. The construction is continued to form an hexagonal grid.

Remark 1: The hexagon in Fig. A8.1b is the projective image of a regular hexagon shown in Fig. A8.1c.

Remark 2: One line has been singled out as the horizon line h. The hexagons of the net are perspective drawings of regular hexagons in which three lines, two opposite edges and the diagonal between the remaining two vertices, are parallel and therefore meet at each of the three vanishing points on the horizon line (see Figs. A8.1b and A8.2).

Construction A8.1: Create one hexagon from this construction.

Constructions A8.2: Extend this hexagon to several other hexagons from the hexagonal net.

A9. Quadrilateral Net

This time we begin with a configuration of four lines as shown in Fig. A9.1a with one line designated as the horizon line h. The three points A, C, D lie on h. This construction will define a fourth point B on h and a quadrilateral along with its two diagonals as follows:

a. Connect all existing points by lines. As Fig. A9.1b shows, two additional lines from C must be added. This defines a quadrilateral with one diagonal intersecting h at D and with vertices labeled $1, 2, 3, 4$.
b. Draw the other diagonal of the quadrilateral and notice that it intersects line h in a fourth point, B.
c. There are pencils of two lines incident to two of the points A, C on h.

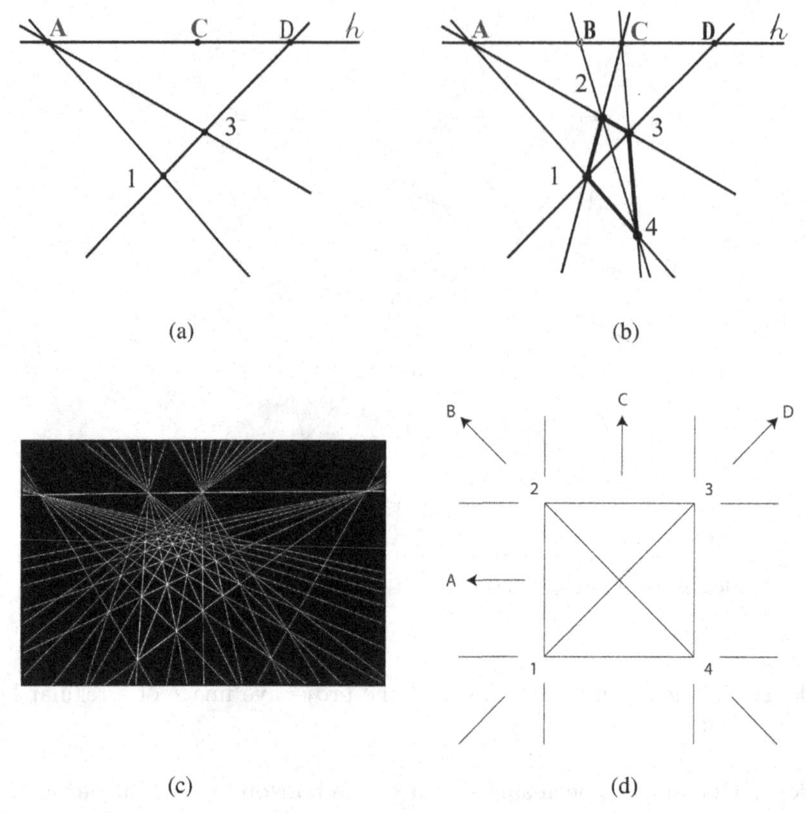

Fig. A9.1. (a) Four lines are given with one line distinguished as the horizon h, and three points A, C, D are specified on h. A fourth point B emerges from the construction; (b) this configuration leads to a construction of a parallelogram in perspective; (c) the configuration defines a grid of parallelograms; (d) a reference square with vertices labeled 1, 2, 3, 4 and points A, B, C, D placed at infinity as an aid to the construction.

Since h is thought to be the horizon line, the two lines in each pencil can be thought of as being parallel edges of the quadrilateral in which case the quadrilateral is the projective image of a parallelogram. A and C are the vanishing points of the parallel edges while B and D are the vanishing points of the diagonals.

d. In Fig. 9.1c, additional lines project from the two pencils to define a net of quadrilaterals where all of the diagonals meet at either B or D.

Remark 1: The two points A, C of h where the edges of the quadrilateral intersect intersperse the two points B, D which go through the diagonals.

The following diagram can be used to clarify this construction:

a. We can consider the parallelogram to be a square as shown in Fig. A9.1d with vertices labeled $1, 2, 3, 4$.
b. The side connecting vertices 1 and 4 will be labeled $(1, 4)$. Then we see that $(1, 4)$ and $(2, 3)$ are parallel edges meeting at A located on the line at infinity (see Fig. A9.1b).
c. Lines $(1, 2)$ and $(3, 4)$ are edges that meet at C, and

d. Lines $(1,3)$ and $(2,4)$ are the diagonals of the rectangle meeting at D and B, respectively.
e. For the rectangle, the vanishing points of the parallel edges and the diagonals are at infinity whereas in perspective they meet on the horizon line h.
f. This analysis can be extended to a grid of rectangles with all the parallel edges meeting at either A or C and the diagonals meeting at B or D.

Figure 9.1d will be useful for helping to visualize the next construction.

Remark 2: The four points A, B, C, D on h are called *harmonic* because their cross ratio equals -1, i.e.,

$$\lambda = \frac{AB}{BC} \div \frac{AD}{DC} = -1.$$

Construction A9.1: From four lines and three points of your own choice, construct a quadrilateral with its diagonals using the procedure described above. Compute the cross-ratio of A,B,C,D and show that it equals -1.

Construction A9.2: Construct a net of at least four quadrilaterals.

A10. Points of the Quadrilateral Net Going to Infinity I

A10.1. Infinity in a projective context

Consider Fig. A10.1, a sequence of lines radiating from a point and intersecting the horizontal line. Notice that as the intersection point on the horizontal line moves from left to right the radiating line rotates counterclockwise until when the point reaches positive infinity, the radiating line is horizontal. Now let the radiating line rotate a slight bit more in a counterclockwise direction. Notice that the point of intersection on the horizontal line has now shifted all the way to the left and begun to come back to its starting point as the line rotates even more. In summary, the point on the horizontal line goes to positive infinity and shifts to negative infinity from which it returns.

In Euclidean geometry, with the exception of parallel lines, all lines intersect at a unique point. Parallel lines, by definition, never meet, and the fifth axiom of Euclidean geometry states that given a line and a point not on the line, there exists exactly one line through the point parallel to the given line (see Chap. 10). In projective geometry, this exception is eliminated, and by an axiom of projective geometry, any two lines share a common point of intersection. If the lines are parallel, that point exists, but it is the infinite point in the direction of the line. A new pair of parallel lines with a

Fig. A10.1. Illustration of the nature of infinity in a projective context.

different orientation meet at a different point located at infinity. All parallel lines of a given orientation meet at the same point at infinity, and the collection of infinite points is said to make up the *line at infinity.* Some times the line at infinity is referred to as the *circle at infinity* since, as we saw above, the left and right infinite points on the line are identified with each other and considered to be the same point. These somewhat abstract ideas are made concrete when the infinite line is mapped to the finite horizon line on the artist's canvas. The points where parallel lines intersect are then clearly visible.

A10.2. A quadrilateral net when the diagonal vanishing point goes to infinity

Reconsider the configuration of the quadrilateral in Fig. A9.1b. What happens to this configuration as point D moves to the right approaching infinity? The diagonal lines through D then approach lines parallel to h. Therefore, one of the diagonals for each quadrilateral of the net must be horizontal as shown in Fig. A10.2. We also find that point B approaches the midpoint of the line segment AC. For the first time we have a precise measurement entering into a projective context.

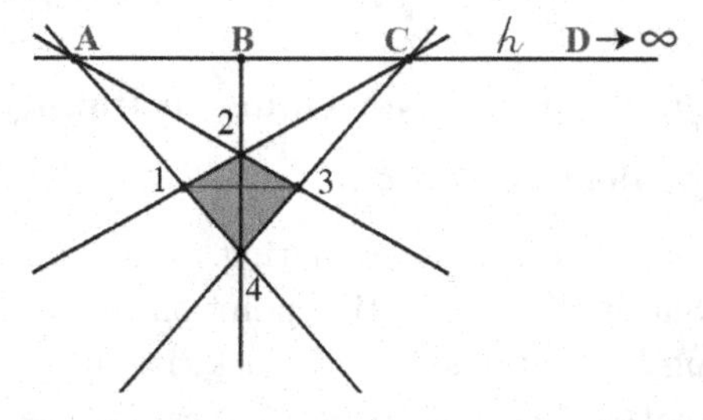

Fig. A10.2. Point D on line h moves to infinity resulting in a parallelogram with a horizontal diagonal.

Construction A10.1: Let D approach infinity and draw a quadrilateral using the previous construction. Show that B is the midpoint of AC.

A11. Points Going to Infinity II

If we let vertex 4 of the schematic rectangle in Fig. A11.1a approach infinity, the lines 14, 34 and 24, since they meet at infinity, must be parallel in the projective parallelogram shown in Fig. A11.1b. To create Fig. A11.1b, we are free to choose two points A, D and three lines: the horizon line h, two lines incident to A, and one line through D as shown in Fig. A11.1b. Although C is no longer given, fixed point 4 is no longer free, but it is located on the line at infinity. The line through D intersects the lines through A at 1 and C at 3. C is now defined by the line 34 drawn parallel to line 14. Now that C is defined, line $1C$ defines point 2 and the other diagonal is the line through point 2 parallel to line 14.

Construction A11.1. Carry out this construction to see what happens as point 4 moves to infinity. Make a construction similar to Fig. 11.1b.

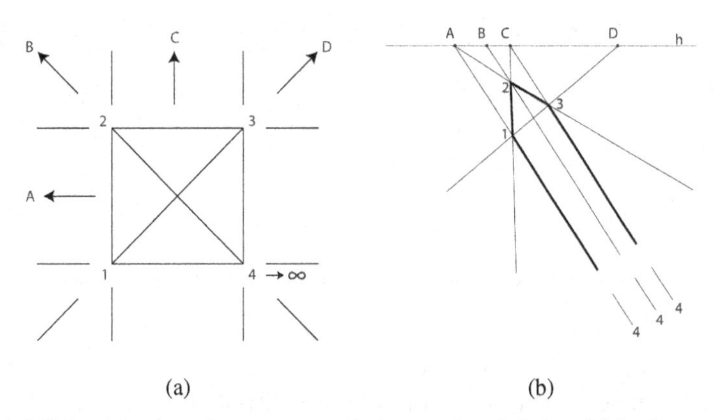

Fig. A11.1. (a) Point 4 in the reference rectangle is moved to infinity; (b) leads to a configuration with an infinite parallelogram.

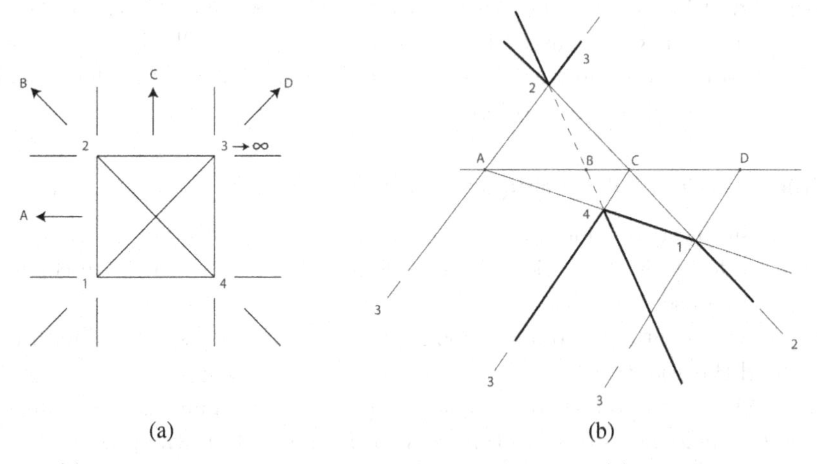

Fig. A11.2. (a) Point 3 in the reference rectangle moves to infinity; (b) leads to a configuration in which point 2 of the parallelogram moves to the other side of h and the parallelogram reconnects at infinity.

Now let us see what happens if point 3 approaches infinity. In the schematic rectangle shown in Fig. A11.2a, lines 13, 23 and 43 must now be parallel in the perspective parallelogram shown in Fig. A11.2b. We are free to choose A, C, D and two lines through A, in addition to point 3 which is now placed at infinity. As the schematic diagram (Fig. A11.2a) shows, one of the lines through A contains the infinite point 3. We are no longer free to choose the diagonal line 13 through D, since this line must be parallel to line $3A$. Where this line intersects the second line through A defines point 1. One of the lines through C must also be parallel to lines $3A$ and 13, and this line intersects one of the lines through A defining point 4. To carry out this construction we begin with line h and two arbitrary lines through A. By the above analysis, points 1, 4 and 3 are now defined, but where is vertex 2? Since the line through A in the direction of 3 and the line through C in the direction of 1 must intersect at 2 (see Fig. A11.2b), we find the somewhat surprising result that vertex 2 must be on the other side of h.

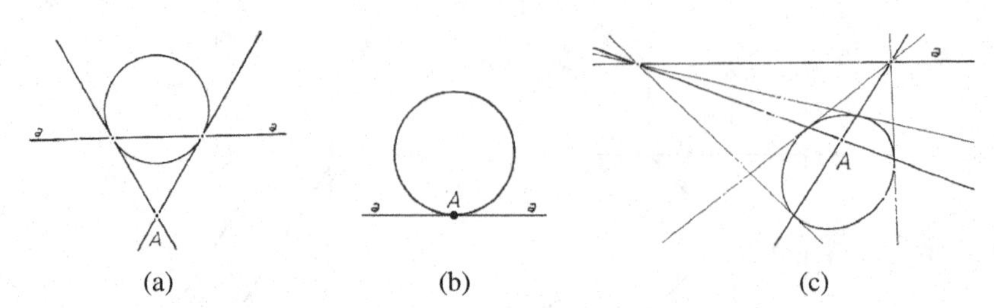

Fig. A12.1. Construction of pole and polar of an ellipse where (a) the pole is exterior to the ellipse; (b) the pole is on the ellipse; (c) the pole is inside the ellipse.

We have highlighted vertex 2 in Fig. A11.2b and see that to get to vertex 2 from 1, we must go though infinity and return from infinity on the other side of h, a remarkable portrait of infinity in a projective context. Since (2,4) is the other diagonal, Fig. A11.2b shows that diagonal going from 2 to 4 through infinity.

Construction A11.2: Carry out this construction to see what happens as point 3 moves to infinity. Make a construction similar to Figs. A11.2b. If you can computerize this, it would be interesting to see an animation showing what happens to the quadrilateral as point 3 recedes to infinity.

A12. Pole and Polar of an Ellipse or Circle

Perhaps the simplest and most elegant construction in projective geometry and its application to design is the notion of the pole and polar of a conic. We consider the conic to be an ellipse or a circle.

The *pole* is a point either outside, inside, or on an ellipse. Associated with the pole is a line called the *polar*.

In Fig. A12.1a, the pole is outside the ellipse or circle. The polar is defined to be the line that is incident to the two tangents to the ellipse from the pole.

In Fig. A12.1b, the pole is a point on the ellipse or circle. The polar is then the line tangent to the ellipse at the pole.

In Fig. A12.1c, the pole is within the ellipse. To find the polar, draw any two lines through the pole. Where each line intersects the ellipse, draw a pair of tangent lines. Draw the line that joins these two pairs of tangent lines at their points of intersection. This is the polar. In fact, the pair of tangent lines for all lines through the pole intersect on the polar as can be seen in Fig. A12.2.

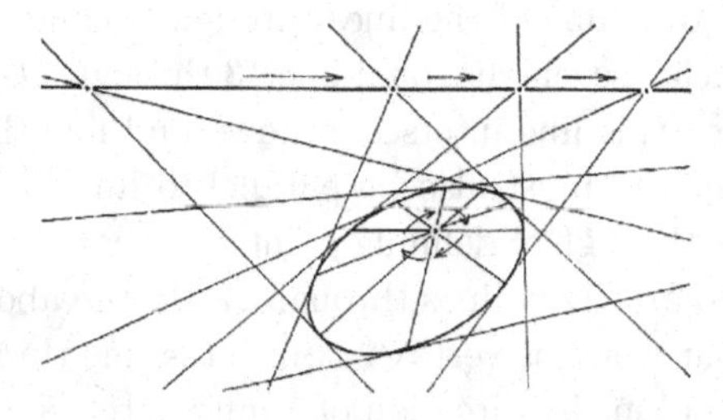

Fig. A12.2. The pair of tangents to each pair of points where the line through the pole intersects a circle or ellipse all meet on the polar corresponding to the pole.

Any line from the pole A intersects the ellipse at S and T and the polar at A' as shown in Fig. A12.3. It turns out that the four points A, S, A', T are harmonic, i.e., their cross-ratio equals -1.

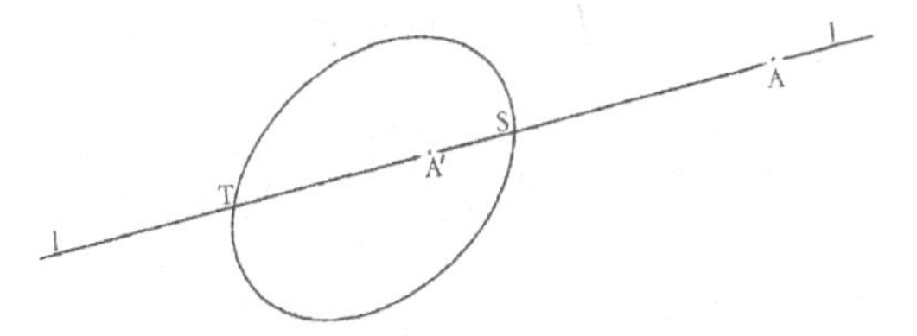

Fig. A12.3. A line from pole A cuts the ellipse at points S and T and the polar at A'. A, S, A', T are harmonic.

Construction A12.1: For an ellipse or for a circle, draw the polar lines for the case where the pole is outside, inside and on the ellipse or circle.

In projective geometry, curves can be defined either pointwise or linewise. For example, Fig. A12.4a shows a set of inner circles defined pointwise. Tangent lines to points on that circle at an evenly spaced sequence of points define a sequence of outer circles in a both linewise and pointwise sense. Fig. A12.4b shows the circles defined linewise.

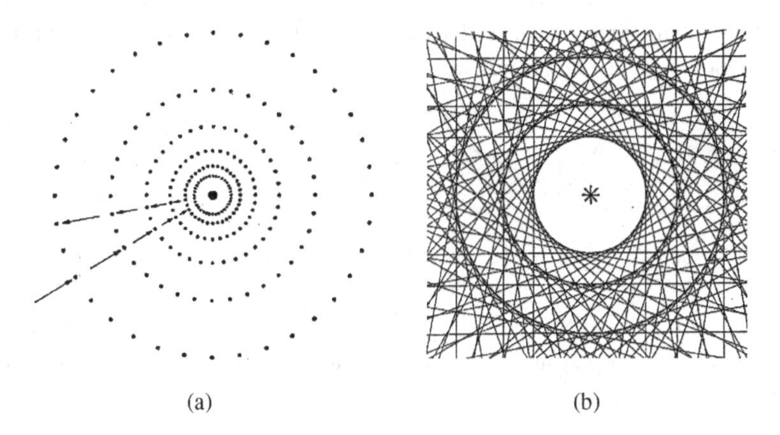

Fig. A12.4. (a) A sequence of circles defined pointwise; (b) A sequence of circle defined pointwise.

A13. Inversion in a Circle

If we consider the relationship between pole and polar in the context of a circle, this leads to a wonderful transformation known as *inversion in a circle* from which many interesting designs can be derived.

1. Consider a circle with radius r and center at O with point A outside of the circle. If A is considered the pole, then line UV is the polar as shown in Fig. A13.1.
2. Where the line through O and A cuts the polar line is where A' is located. We can say that A maps to A' under *inversion in a circle* and vice versa.

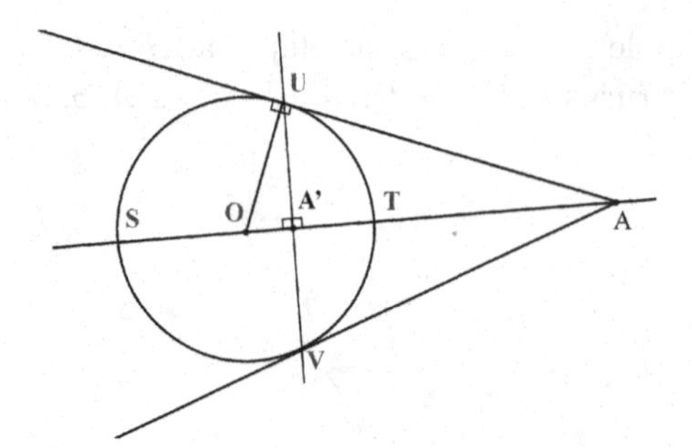

Fig. A13.1. Diagram illustrating inversion in a circle.

3. Since triangle UOA' is similar to UOA (why?) it follows that,

$$\frac{OA'}{r} = \frac{r}{OA}. \tag{A13.1}$$

If the radius is considered to be one unit, then OA and OA' are inverse distances from the center of the circle, i.e.,

$$OA' = \frac{1}{OA}. \tag{A13.2}$$

4. It is easy to find a polar when the pole A lies within the circle. On the line OA, locate A' by using either Eqs. A13.1 or A13.2. The polar line will be perpendicular to OA' at point A' as shown in Fig. A13.2.

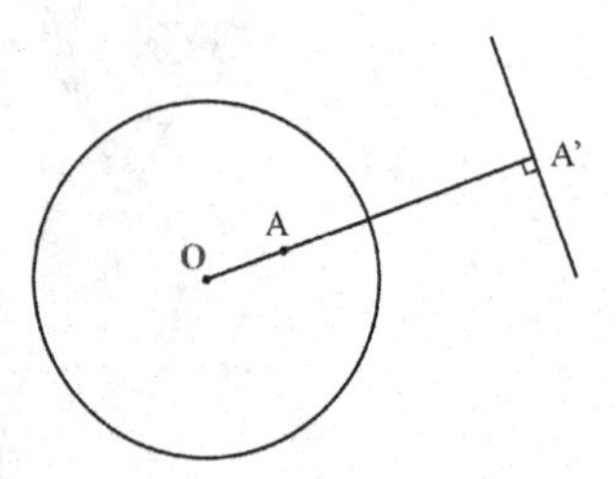

Fig. A13.2. A and A' are inverse with respect to the circle. If A is the pole then the polar is perpendicular to A'.

5. A circle of points (poles) are taken within the circle in Fig. A13.3. Each point has a polar constructed as in Step 4. Notice that the envelope of lines maps out the arc of a circle exterior to the circle. Note: The envelope of a curve is the set of tangent lines to each point on the curve.
6. Next consider any curve within a unit circle. How can we use Eq. A13.2 to map this curve to the exterior of the circle? We will demonstrate this mapping for three points A, B and C on the inner curve (see Fig. A13.4). We will be able to find the envelope of the outer curve by determining the polars a, b, c corresponding to points A, B, C as we did in Step 5.
7. To find the polars a, b, c first locate A', B', C' using Eq. A13.1. The polar lines will be at right angles to lines OA', OB', OC'. If enough points are taken on the inner curve, the outer curve will be accurately defined by its envelope. To find the polar

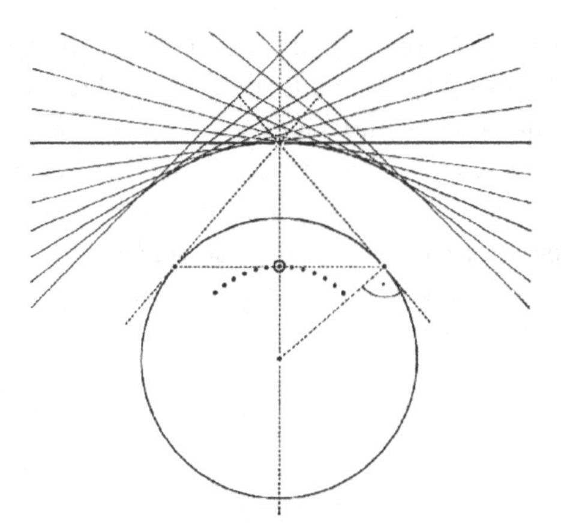

Fig. A13.3. Points on a circle inside the reference circle map to an envelope of lines exterior to the reference circle.

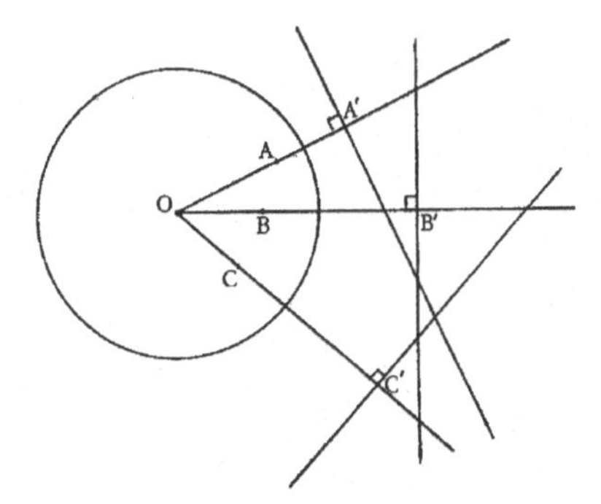

Fig. A13.4. Detail of three points inside reference circle mapping to three lines outside.

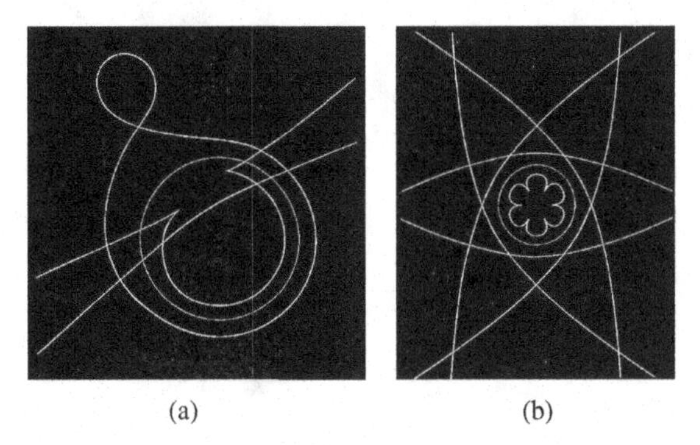

Fig. A13.5. (a), (b) Example of a two designs in which a curve internal to the circle of inversion is mapped to an outside curve.

envelope, we only have to move around this curve with a set-square, keeping one arm of the set-square through O, and the right angle on the curve. In this way we can draw as many tangents to our polar curve as we wish or find the time for.

8. Two lovely inversions of curves are shown in Figs. A13.5a and A13.5b.

9. In this inversion, every point on the circumference of the circle of inversion transforms to itself, i.e., remains invariant; points in the interior of the circle are transformed to external points and vice versa; convex figures are transformed to concave and vice versa; lines tangent to the circle of inversion are transformed to circles touching the circle of inversion and containing the center O of the circle of inversion and vice versa; circles containing the center O transform to lines exterior to the circle; and angles between lines are preserved under this transformation.

APPENDIX B

GLOSSARY

Altitude — A line drawn from the vertex of a triangle perpendicular to the opposite side.

Acute angle — An angle less than 90 deg.

Acute triangle — A triangle with all acute angles.

Alternate interior angles — A pair of angles that are equal when a transversal intersects a pair of parallel lines.

Area — A measure of the content of a plane figure in terms of the number of unit squares that fit within it.

Associative law — A law of algebra stating that $(ab)c = a(bc)$.

Axiom — An apriori premise assumed to be true that initiates a mathematical system and for which all true statements within the system are consequences.

Bipartite graph — A graph composed of two sets of vertices in which vertices from the first set may be connected to vertices from the second, but not to vertices within its own set.

Central angle — The angle between a pair of radii subtending an arc of a circle.

Centroid — The balance point of an object or geometric figure with uniform density.

Chord — A line segment drawn between two points on the circumference of a circle.

Circumcenter — The meeting point of the perpendicular bisectors of a triangle.

Circumscribing circle — A circle about the circumcenter of a triangle that goes through the vertices of the triangle.

Complementary angles — A pair of equal angles the result of cutting two parallel lines by a transversal.

Commutative law — A law of algebra stating that $ab = ba$.

Congruent triangles — A pair of triangles with equal sides and equal angles.

Conic section — When a double cone is intersected by a plane, the line of intersection is either a circle, ellipse, parabola, or hyperbola.

Conclusion — The consequence of the hypothesis of a theorem.

Connected graph — A graph in which there is a path connecting each pair of vertices.

Converse — A statement formed by interchanging the hypothesis and conclusion in the expression of a theorem.

Convex curve — A closed curve in which a line drawn between any pair of points inside the curve lies entirely within the curve.

Cross-ratio — Given four points A, B, C, D on a line, the cross-ration, $\lambda = \frac{AB/BC}{AD/DC}$. The cross-ratio is invariant under projections from a point or a series of points.

Cycle — A path leading from a point of a graph back to itself.

Determinant — A procedure that generates a number given an $n \times n$ matrix.

Dot product — The components of two vectors are multiplied and the results are added, i.e., if $\vec{v_1} = (a_1, a_2)$ and $\vec{v_2} = (b_1, b_2)$ the dot product is $\vec{v_1} \bullet \vec{v_2} = a_1b_1 + a_2b_2$.

Duality — In plane projective geometry, any true statement remains true when point is replaced by line and line by point. Such statement pairs are called dual.

Dynamic symmetry — A procedure developed by Jay Hambridge that divides a rectangle into another rectangle with the same proportions and a leftover rectangle called a gnomon.

Edge — A line drawn between two vertices of a graph. The line need not be straight.

Ellipse — An oval shaped member of the family of conic sections formed by the intersection of a cone by a plane slicing obliquely through the cone.

Equilateral triangle — A triangle with three equal sides and three angles of 60 deg.

Euler line — A line containing the circumcenter, orthocenter and centroid of a triangle.

Fractal — A curve that is self-similar at a variety of scales.

Frieze symmetry — The set of seven symmetry classes along a line.

Geometric sequence — A sequence in which the ratio of successive terms is constant.

Glide reflection — An isometry obtained by translating a figure and reflecting it in a mirror line in the direction of the translation (like footprints in the snow).

Gnomon — For a rectangle of given proportions, the gnomon is a rectangle that, when added, results in an enlarged rectangle of the same proportion.

Golden section — An irrational proportion gotten by dividing a line segment in such a way that the ratio of the whole is to the larger part as the larger is to the smaller. The proportion is: $1 : \frac{1+\sqrt{5}}{2}$.

Graph — A collection of vertices and edges with information as to how the vertices pair to form the edges.

Group — A set of objects, a well-defined operation, and four rules determining how the objects can be combined as a result of the operations.

Hexagon — A polygon with six sides.

Horizon line — The transformation of the line at infinity in a scene to a horizontal line drawn on an artist's canvas.

Hyperbola — A curve formed by intersection of a double cone with a plane that cuts through both cones.

Hypothesis — A statement assumed to be true that leads to the conclusion of a theorem.

Identity transformation — A transformation that transforms each point to itself.

Incenter — The meeting point of the angle bisectors of a triangle.

Inscribed circle — A circle about the incenter tangent to the sides of a triangle.

Inscribed angle — An angle whose vertex lies on the circumference of a circle intercepting an arc of the circle.

Inverse — Given a transformation, the inverse is a second transformation whose product with the first results in the identity.

Isosceles triangle — A triangle with two equal sides.

Isometry — A transformation that preserves the distance between points.

Kaleidoscope symmetry — A set of symmetries about a point carried out by the reflections in a pair of mirrors intersecting at the point in a set of discrete angles $\theta = \frac{180}{n}$ for integer n.

Lemma — A theorem that is used to prove other theorems.

Line — An undefined quantity of geometry that is made up of a set of points extending to infinity in both directions.

Line segment — A line drawn between two finite points.

Line chopper — A geometric construction of equally spaced lines perpendicular to a given line that enables lengths to be subdivided into any number of equal line segments.

Logarithmic spiral — The only smooth curve that is self-similar at all scales. As its central angle doubles its radius squares.

Matrix — An ordered rectangular array of numbers.

Median — A line drawn from a vertex of a triangle to the midpoint of the opposite side.

Obtuse angle — An angle larger than 90 deg.

Obtuse triangle — A triangle that possesses an angle larger than 90 deg.

Octagon — A polygon with eight sides.

Orthocenter — The meeting points of the altitudes of a triangle.

Pappus line — A third line that is the result of joining three points on each of two lines.

Parabola — The intersection of a cone with a plane oriented parallel to the side of the double-cone.

Parallel lines — A pair of lines that never meet in the finite plane.

Parallelogram — A polygon with opposite sides equal and parallel.

Pascal line — The line formed by the intersection of the opposite sides of a hexagon that has been inscribed in a circle or ellipse.

Path through a graph — A sequence of vertices and edges leading from one point to another in a graph.

Pencil of lines — A collection of lines emanating from a point of the plane.

Pentagon — A polygon with five sides.

Perimeter — A measure of the length of the boundary curve surrounding a two dimensional region.

Perpendicular bisector — A line that intersects a line segment at right angles and bisects it.

Pick's law — A simple equation that can be used to calculate the area of a polygon drawn on a lattice of points, $A = I + C/2 - 1$, where A is the area, I is the number of lattice points within the polygon, and C is the number of lattice points on the perimeter of the polygon.

Pixel — A mark generated by a computer to simulate a point.

Poincare plane — A circle that serves as a model for one type of non-Euclidean geometry. The circle represents the points at infinity while the arc of a circle meeting this circle at right angles represents a line.

Point — An undefined object in Euclidean geometry that represents a location in the plane.

Polar — A line in the plane generated by a point known as a pole with respect to a circle or ellipse. The line may intersect, be tangent to, or lie outside of the circle.

Pole — A point that lies outside, on the circumference of, or inside a circle or ellipse and generates a line known as the polar.

Polygon — A cycle of lines and vertices.

Postulate — An apriori statement assumed to be true about a mathematical system from which all true statements within the system, i.e., theorems are derived.

Pythagorean triple — Three integers that represent the side lengths of a right triangle.

Quadrilateral — A four-sided polygon.

Ray — A half-line originating from a point.

Rectangle — A polygon with opposite edges equal having all right angles.

Reflection — An isometry generated by a mirror line. A point on one side of the line is mapped to a point equidistant on the other side of the mirror.

Regular polygon — A polygon with equal sides and angles.

Rhombus — A parallelogram with all sides equal.

Rotation — An isometry in which a point is rotated by an angle about a fixed point or center.

Sacred cut — An arc of a circle generated by a corner of a square through the midpoint of the square which intersects a side of the square in the ratio $1 : \theta$ where $\theta = 1 + \sqrt{2}$.

Scalene triangle — A triangle with unequal sides and angles.

Silver mean — The irrational number $1 + \sqrt{2} = \theta$.

Similarity — Two figures are similar if one is a magnification or contraction of the other.

Slope — Rise/run of a line.

Square — A four-sided polygon with all equal sides and all right angles.

Subgraph — A graph all of whose vertices correspond to a second graph while the second graph can have additional edges.

Supplementary angles — A pair of adjacent angles that sum to 180 deg.

Theorem — A true statement that can be derived from a set of postulates, previously proved theorems, definitions, hypotheses, and laws of logic.

Translation — An isometry in which points are transformed by moving them according to a given vector.

Transversal — A line that cuts a pair of parallel lines obliquely.

Trapezoid — A quadrilateral with one pair of parallel sides.

Tree graph — A connected graph with no cycles.

Vector — A quantity with magnitude and direction but independent of the starting point.

Vertical angles — The opposite angles at the point of intersection of a pair of lines. Vertical angles are equal.

Vesica Pisces — A construction in which a circle is drawn, and a second circle with the same radius is drawn centered about an arbitrary point on the circumference of the first.

Voronoi domain — Given a set of points in the plane, the Voronoi domain of a point is all of the points of the plane nearer to that point than the other points in the set.

APPENDIX C

ESSAYS

Respond to the following statements with two-page essays.

1. My experience with geometry:
   Describe your first experiences and encounters with geometry and number both in and out of school.
2. The difference between art and design:
   I once made a presentation of some of the best student examples that I had of Amish quilt designs. An artist in the audience commented that these designs were quite pleasant but they were not art, although this is a first step to creating art. To make them into works of art, you would have to eliminate their austere regularity and symmetry while retaining remnants of their sense of order. In doing this, the designer would be making a personal statement. However, the step of creating total order is an essential first step before engaging in the process of de-symmetrizing the design or rule breaking.
   Describe what you feel is the difference between art and design.
3. Transformation and Geometry:
   Transformation lies at the base of how people learn. For example, children learn about their world by manipulating or transforming the objects around them. Through the use of metaphor, poets and artists map the world of ideas onto their work, enabling the rest of us to sharpen our understanding of these ideas by seeing them in a different light.
   Comment on this statement. Since mathematics deals primarily with transformation, state your opinion about whether mathematics can serve as a useful metaphor for architecture and design. Cite ideas that you have been exposed to in your geometry course that may be applicable to architecture or design.
4. Order and Symmetry:
   In his paper entitled "Perception and Modular Coordination", Christopher Alexander suggests that we enjoy symmetric themes in design because our minds recoil at chaos but are put at ease by the repetition of a simple motif. We like to see things that look familiar, that we have seen before, and structure and order in art and

architecture help us to feel secure and comfortable with our surroundings. On the other hand, people react adversely to mindless, monotonous order. To an extent, it is the job of the artist, architect, and designer to supply, through their work, a solution to the problem of satisfying the needs of people for both order and novelty.

Comment on this statement. Do you agree or disagree? The snowflake patterns and Frieze patterns and other symmetric patterns that you have been exposed to this semester certainly satisfy the criterion of design based on order and repetition. Is it also capable of producing designs interesting enough to appeal to our need for surprise and novelty?

5. It is stated in the Kaballah, a book of Jewish mysticism, that there are actually two Bibles or Torahs handed to Man by God: One of the written words and the other, made up of the space between the words. Musicians are also aware that the space between the performed notes are as important as the notes themselves.

   Give your opinion as to whether the space left empty within a design has equal importance to the space that is filled.

BIBLIOGRAPHY

[Bru] Brunes, T. (1967) *The Secrets of Ancient Geometry — And its Use* (Rhodos, Copenhagen)

[Ech] Echols, M. Private communication

[EdwL] Edwards, L (1985) *Projective Geometry* (Rudolf Steiner Institute, Phoenixville, Pa.)

[EdwE] Edwards, E. *Pattern and Design with Dynamic Symmetry* (Dover, New York)

[Ern] Ernst, B. (1976) *The Magic Mirror of M.C. Escher* (Ballantine Books, New York)

[Flo] Flower of Life http://en.wikipedia.org/wiki/Flower_ of_ Life

[Fra] Frantz, M. and Crannell, A. *Viewpoints: Mathematical Perspective and Fractals in Art*

[Gar] Gardner, M. (1978) *Aha, Insight* (Scientific American)

[Ham] Hambridge, J. (1967) *The Fundamental Principles of Dynamic Symmetry* (Dover, New York)

[Kap1] Kappraff, J. (1990) Connections *The Geometric Bridge between Art and Science*, 2nd Edn. (World Scientific, Singapore)

[Kap2] Kappraff, J. (2001) *Beyond Measure: A Guided Tour through Nature, Myth, and Number* (World Scientific, Singapore)

[Liv] Livio, M (2002) *The Golden Ratio: The Story of Phi, the World's Most Astonishing Number* (Broadway books, New York)

[Man] Mandelbrot, B.B. and Frame, M.L. (2002) Fractals, Graphics, and Mathematics Education (Mathematics Association of America)

[Mey] Meyer, W. (2006) *Roads to Geometry* (Academic Press, San Diego)

[Mum] Mumford, D, Series, C., Wright, D. (2002) *Indras Pearls: The Vision of Felix Klekin* (Cambridge Press, New York)

[Ols] Olsen, S. (2006) *The Golden Section* (Walker and Company, New York)

[Pei] Peitgen, H., *et al.* (1991) *Fractals for the Classroom: Strategic Activities* Volume One (Springer Science + Business Media, New York)

[Rad] Radovic, L. *ni-rs.academia.edu/* ***LjiljanaRadovic***

[Spe-B] Spencer-Brown, G. (1969) *Laws of Form* (George Allen and Unwin Ltd., London)

[Wal] Wallace, E.C. and West, S.F.(1998) *Roads to Geometry*, 2nd Ed. (Prentice Hall, Upper Saddle River, NJ)

[Wat] Watts, D.J. and Watts, C. (Dec. 1986) "AA Roman Apartment Complex", Sci. Am., Vol 255, No. 6, 132–140.

[Whi] Whicher, O. (1971) *Projective Geometry: Creative Polarities in Space and Time* (Rudolf Steiner Press, London)

INDEX

Printed in the United States
By Bookmasters